Technology and the Blue Economy

Technology and the Blue Economy

From autonomous shipping to big data

Nick Lambert
Jonathan Turner
Andy Hamflett

Publisher's note

Every possible effort has been made to ensure that the information contained in this book is accurate at the time of going to press, and the publishers and authors cannot accept responsibility for any errors or omissions, however caused. No responsibility for loss or damage occasioned to any person acting, or refraining from action, as a result of the material in this publication can be accepted by the publisher or the authors.

First published in Great Britain and the United States in 2019 by Kogan Page Limited

Apart from any fair dealing for the purposes of research or private study, or criticism or review, as permitted under the Copyright, Designs and Patents Act 1988, this publication may only be reproduced, stored or transmitted, in any form or by any means, with the prior permission in writing of the publishers, or in the case of reprographic reproduction in accordance with the terms and licences issued by the CLA. Enquiries concerning reproduction outside these terms should be sent to the publishers at the undermentioned addresses:

2nd Floor, 45 Gee Street	122 W 27th St, 10th Floor	4737/23 Ansari Road
London	New York, NY 10001	Daryaganj
EC1V 3RS	USA	New Delhi 110002
United Kingdom		India

www.koganpage.com

© Nick Lambert, Jonathan Turner and Andy Hamflett 2019

The right of Nick Lambert, Jonathan Turner and Andy Hamflett to be identified as the authors of this work has been asserted by them in accordance with the Copyright, Designs and Patents Act 1988.

ISBNs

Hardback	978 1 78966 0227
Paperback	978 0 74948 3951
Ebook	978 0 74948 3968

British Library Cataloguing-in-Publication Data

A CIP record for this book is available from the British Library.

Library of Congress Cataloging-in-Publication Data

A CIP record is available from the Library of Congress.

Typeset by Integra Software Services, Pondicherry
Print production managed by Jellyfish
Printed and bound by CPI Group (UK) Ltd, Croydon CR0 4YY

CONTENTS

Preface x

01 An introduction to the Blue Economy 1

An ocean of opportunity 1
Embracing complexity 2
References 5

02 Shipping 6

Navigating an unprecedented sea of data 6
Playing 'spoof' 10
Rising to the cyber challenge 12
Tackling the energy challenge 13
Plain sailing? 16
From reaction to prediction 17
The power of sharing 23
Conclusion 24
References 28

03 Ports and harbours 31

Leading the global charge against harmful emissions 31
Ports in a storm 32
Safe berth 35
Something in the air 36
Taking down the particulates 40
Setting the scene for innovation 42
Conclusion 44
References 45

04 Offshore renewables 48

Channelling the power of the oceans 48
A climate of fear 50
Winds of change 55
Floating a new idea 58
Maintenance from afar 59
Continuous operation 62
'On land' at sea 63
Potential downsides 65
Cross-cutting value 66
Conclusion 67
References 69

05 The cruise industry 71

Pushing the digital boundaries of the customer experience 71
Greener and cleaner 74
Coming in from the cold 75
Back to the future 76
Floating on the digital revolution 77
Getting personal 80
Safe and secure 82
Okay, Zoe? 84
Reaching for the stars 85
Conclusion 86
References 88

06 Maritime surveillance 89

Maintaining eyes on the sea 89
SAR-struck 93
Dark matters 97
Oil on the water 99
Digital dodging 100
Robocams 102
Quality, quantity and safety of data 103
Conclusion 106
References 108

07 Aquaculture 109

Meeting the world's food needs in an environmentally
   sound manner 109
The global race 112
In full bloom 115
Communication breakdown 117
Net effect 119
Imprisoned in steel 121
Conclusion 122
References 125

08 Hydrography and bathymetry 129

Bringing clarity to the ocean floor 129
One data set, multiple uses 131
Actionable intelligence 132
Ports, boundaries and offshore developments 133
From theory to action 135
Order from chaos 136
Safety blanket 138
'Irmageddon' 139
A light in the dark 141
Setting the future course 143
Conclusion 144
References 146

09 Ocean conservation 147

Protecting and preserving the future of the seas and oceans 147
A broader challenge 149
Micro waves 151
Harnessing the social network 153
Going with the flow 156
Power to (and from) the people 158
Sounding off 161
Pushing the boundaries 162
Reap what you sow 165

Conclusion 166
References 169

10 Sustainable fisheries 173

Casting a digital net to help feed the world 173
A light in the blue 176
Providing the framework 177
A multi-faceted problem 179
Airborne AI 182
The open data challenge 189
Conclusion 192
References 193

11 Subsea monitoring 194

Shining a light on the mysteries of the deep 194
Advanced AUVs 196
Establishing the search rhythm 198
Continual challenge 199
Needles in a watery haystack 201
All eyes on flight MH370 202
Raising the profile of dangerous wrecks 205
New possibilities 206
The 'system of systems' 207
Safeguarding the world's data 209
The fragility of the connected world 211
Getting power back to shore 213
Conclusion 215
References 217

12 Safety of life at sea 218

Protecting human life in the harshest of environments 218
VR, wearables and robots to the rescue 221
No safety net 223
Forewarned is forearmed 226
Caught in the net 227

Cruise control? 228
Proof positive 232
Funding the future 233
Invest to save 234
Conclusion 235
References 238

13 Conclusion 239

Successfully navigating a sea of opportunity 239
The keys to successful product development in the Blue Economy 240
Be wanted: Meet a need 241
Be user-led: Learn and pivot 243
Be sustainable: Develop the business model 245
Be patient and resilient: Customer take-up might take time 248
Addressing the Blue Economy funding network 250
Incubation and acceleration 253
The key enabler: A shift in investment priorities? 254
The greatest gifts 256
Reference 258

Index 259

PREFACE

Interest in the seas and oceans is enjoying something of a boom time in the public consciousness at the moment. This heightened awareness arises from a convergence of dark messaging about the effects of global warming on ocean communities and ecosystems, recent gut-wrenching televisual images about environmental challenges (all easily shareable across social media channels) and, more positively, a growing understanding of the fact that the ocean economy has huge potential and yet is only starting to be fully explored.

This book was inspired and deeply informed by the ongoing investigations and activities of Blue Economy solutions company NLA International (NLAI). The company was formed by this book's three authors, and for the past few years we have combined our experiences in the naval, hydrographic, business and NGO sectors to explore and promote innovative activity within the Blue Economy. Each of the NLAI directors has in various ways worked with, relied on and promoted technology and, especially, data in their previous careers, and it has been fascinating to apply cross-cutting insights and experiences in the ocean sector as we work intensively with companies attempting to effect change with cutting edge new technologies.

We thought the general public – and those within a variety of Blue Economy industries – would benefit from a snapshot of the amazing technologies that are emerging to help protect, preserve and benefit from the seas and oceans. With the public focus so heavily (and in many ways usefully) locked on some of the major challenges facing the planet's oceans, it can be easy to overlook the many reasons to be excited, especially when considering some of the potentially game-changing technologies that are being developed by Blue Economy innovators all over the world.

For example, sea-going vessels have of course always been operated by the people on board – or at least they were until recently. Nowadays, there is an unstoppable rise in marine autonomous systems, remotely operated using terrestrial radio networks and satellite connectivity, and full-scale autonomous ships are becoming a reality. While truly autonomous ships are still relatively rare, smart connected ships are already rapidly increasing in numbers. Fully integrated ships systems (bridge, propulsion and ancillary) will be continuously connected with shore authorities through satellite communications and advanced machine learning software, transforming the global supply chain and driving news levels of efficiency into all kinds of maritime operations.

Within the aquaculture sector (the continued growth of which is absolutely essential to meet the food needs of a rising global population), we are seeing the early adoption of preventative maintenance practices more often witnessed in the aviation sector, and a surge in enthusiasm for remote sensing and machine learning.

Further out to sea, crowdsourced situational awareness data is taking on new prominence in supporting fisheries management organizations. For example, empowering local fisherfolk with accessible, resilient technology that can correlate location- and time-stamped data with a photograph taken by the fisherfolk is a new piece of the maritime domain awareness jigsaw. The images are combined in a data collation hub with other sources such as automatic identification system and vessel monitoring system vessel tracking services, Earth observation data and reports from governance vessels. Thus, empowered fisherfolk become the champions of good fisheries management, protecting artisanal livelihoods and supporting a sustainable marine ecosystem.

When the area of seaspace of a country's Exclusive Economic Zone (EEZ) is considered alongside the land, the world's perspective of a 'big country' changes. What were once small islands are now being considered as large ocean nations, rich in natural resources and precious ecosystems. Marine Protected Areas are one way to manage these vast areas of ocean and the evolution of space-enabled technologies allows us to monitor these areas more than ever before. Such capabilities will play a key role in sustaining the ocean environment and the people and economies that depend upon it.

Such use of space-based assets has the potential to be a real game-changer across all sectors of the Blue Economy. The phenomenon of 'sea blindness' has persisted for as long as people have been at sea; the ocean can be an inhospitable, lonely place without radio connectivity or the ability to see further than the visual horizon. Although radar was a game-changer, it too only extends situational awareness as far as the radar horizon. Space therefore has the real potential to truly transform maritime operations: giving the mariner more connectivity than ever before; providing incredibly accurate position, navigation and timing information; and broadening horizons with high-definition optical and radar images. Radio frequency detection and analysis will continue that transformation, offering the fourth data stream from space. Coupled with increasingly sophisticated terrestrial services, these space-based technologies herald an era of 'sea vision' in which we will soon see all that we need to support the Blue Economy.

So, everyone can hopefully draw some comfort from the astonishing array of new technologies and techniques being developed and in many cases already being applied in all corners of the ocean economy.

For those already working within related sectors, we also hope that this book will provide a useful overview of those technologies breaking through, and provide an easy way in to see how similar (and differing) approaches are gaining traction in tangential areas. Which ones are most sought after by customers? And what are the challenges to the integration of such new approaches within some systems that may be rooted in centuries of practice?

Finally, for innovators working heroically to bring new technologies into use within the Blue Economy, we hope that this book provides a useful framework within which to set their own endeavours. It highlights common themes across several Blue Economy sectors (it can be easy to just be consumed by your own), and shines a torch on common themes and, in some cases, frustrations.

While this book necessarily touches on many broader themes to set the scene, some of which have been developing over decades or even centuries, the majority of the content focuses particularly on how governments, private companies, individuals and communities are investing in, supporting and utilizing new technological approaches to deliver greater benefit in and on the world's seas and oceans.

There were many potential ways to focus in on this issue. Rather than grouping technologies together (eg a chapter on autonomous vessels, one on remote sensing, etc), we felt it would be more useful to focus on market segments, or subsectors, of the Blue Economy. Not only does this make it easier for the reader with a particular sector interest, it also underlines how individual technologies are finding uses across sectors, often in quite different ways. Autonomous vessels, for example, are finding commercial, academic and charitable uses in fields as diverse as oil and gas, security, oceanography, shipping, aquaculture and more.

Similarly, some cross-cutting themes also appear at several times throughout the chapters. Environmental concerns, for example, are very close to the authors' collective hearts, but it is folly to focus on conservation only as a separate, discrete segment; rather, it must also be considered as a vital conversation that is happening right across Blue Economy sectors, both by those driving forward positive change, and by those attempting to resist its urgency in favour of short-term gain. Thankfully, the numbers in that latter group appear to be declining, though there is still so much to do.

When we started out, we broadly knew which areas of the Blue Economy we wished to focus on, but we did not have any hard-and-fast messages to convey, nor did we wish to highlight any particular technologies. Instead, we adopted an intensive watching brief; we tracked scores of related search terms over several months, then took a deep dive into what was happening most prevalently in each sector. This enabled us to take a very up-to-date, news-driven look at the most vibrant topics in our areas of interest. For this reason, most of the technological advances featured in this book have happened within the past 12 months, though many stand on the shoulders of previous breakthroughs. By whittling down thousands of news stories, we were able to focus in on the key technology stories in each sector, and turn our attention to individual technologies or companies that best exemplified the opportunities and challenges therein.

We believe passionately in the potential of modern technology to take the Blue Economy to new heights. Big leaps in performance require big leaps in innovation. With the bar set high, the chances for failure also increase. We're very fortunate to be able to work with a real cross-section of Blue Economy leaders. That includes the very best operators, experts, scientists, creatives and innovators, and includes all forms of organization from tech start-ups, to corporate multinationals and governments, big and small. We feel it's very important to view progress and challenges in the round, and not from any one particular point of view. It is with this balanced viewpoint that we approached the content. This book is not sequential. Each chapter stands alone, though some will reference other chapters that may contain greater context or more detailed explanations of specific technologies.

Finally, even though the Blue Economy is extremely broad, from time to time we also took a little time out to reference technology developments that we were aware of in other sectors – either to highlight shared challenges, or to illustrate how progress is being made that could be mirrored.

We are hugely grateful to everyone we have learned from in the course of writing this book, and to all at Kogan Page for their great support. We hope you enjoy reading the examples and insights we share in this book.

01

An introduction to the Blue Economy

An ocean of opportunity

Seventy per cent of the planet is covered by water. Eighty per cent of global economic trade is transported by sea (United Nations Conference on Trade and Development, 2018). Our seas and oceans are an abundant blend of environmental, ecological, scientific, cultural and mineral riches that are inextricably linked with the sustainability of life, food security and the global economy.

In its seminal work, *The Ocean Economy in 2030* (OECD, 2016), the Organization for Economic Co-operation and Development defines two categories of ocean-based industries – established and emerging. Established Blue Economy industries – those that have reached a certain level of maturity – include industrial capture fisheries, shipping and maritime and coastal tourism, while emerging sectors include industrial marine aquaculture, offshore renewables and maritime safety and surveillance. All are addressed here. When it comes to the potential use of new technologies, both established and emerging industries face their own unique challenges as well as some that are shared.

The traditional view of the scope of the maritime industry, taking in such sectors as shipping and fishing, has been overtaken by the advent in recent years of the Blue Economy. Definitions of the Blue Economy vary widely. The World Wide Fund for Nature (WWF) have typically focused on the sustainable elements, but they recognize that, for others, it simply refers to any economic activity in the maritime sector, whether sustainable or not. The term therefore combines the traditional maritime sectors of shipping, fishing and oil and gas with the newer industries such as aquaculture, renewable energy, biotechnology and coastal tourism. In its broadest terms, then, the Blue Economy aims to protect, preserve and derive greater environmental, economic and social benefit from the seas and oceans.

Taking aquaculture as an example of the new industries, it will be vital in feeding the growing world population, which according to the United Nations will rise from 7.3 billion in 2015 to 8.5 billion in 2030, 9.7 billion in 2050 and 11.2 billion in 2100. Future growth in fish production and consumption is expected to come from aquaculture, which is an efficient protein production method (in comparison to terrestrial livestock farming), and can produce traceable, high-quality, healthy seafood in large volumes.

In 2015, WWF assessed the value of key ocean assets at over US$24 trillion, with two-thirds of that based on assets that require healthy, productive oceans. In fact, based on gross marine product, the ocean may be considered the world's seventh largest economy. According to the European Commission, the Blue Economy contributes just under €500 billion per year to the European economy, supporting 5.4 million jobs.

The total global value of the Blue Economy is predicted to rise to US$3 trillion by 2030, and employ some 40 million people. The OECD predicts that the industry segments that are predicted to achieve greatest growth in the coming years are: offshore wind (estimated to grow 206 per cent between 2010 and 2030); fish processing (a comparatively modest 6 per cent); industrial marine aquaculture (5 per cent); port activities (4.5 per cent); and industrial capture fisheries (4 per cent) (OECD, 2016).

Embracing complexity

It is not until you start delving deeper into each of the Blue Economy sectors – both emerging and established – that you begin really to understand the depth, breadth, complexity and diversity of issues that cut across the Blue Economy in all its forms.

Looking ahead across the next decade, the world in 2030 will be shaped by powerful global trends, all of which have an impact on the Blue Economy. Population growth and rising urbanization will lead to increasing difficulties in providing adequate amounts and quality of food and water and increasing expectations with regard to health, safety, security and environmental impact. Significant economic trends are the increasing gross domestic product (GDP) share of developing countries, a continued growth of energy consumption and a growing volume of trade with changing patterns. The maritime sector will be significantly affected by these trends, and these headlines have been categorized by the Community of European Shipyards (CESA, 2016) as involving issues as broad as extreme weather due to climate change, food and water supply challenges, waterborne trade growth and population growth.

To face such challenges and opportunities in the coming years, maritime stakeholders need to maintain a competitive edge in the market and are on the lookout for innovative ways to achieve this. The potential for innovation appears to be deepening (as evidenced in the greater emphasis on such areas as digital ships), and maritime surveillance capabilities can play a broad and supporting role across the piece.

Of significant interest in recent years, the Blue Economy provides growing opportunities for satellite technologies in services and solutions that enable sustainable growth within the maritime environment, also known as 'Blue Growth'. The essential components for delivering Blue Growth can be viewed as: marine knowledge to improve access to information about the sea; maritime spatial planning to ensure an efficient and sustainable management of activities at sea; and integrated maritime surveillance to give authorities a better picture of what is happening at sea.

Consideration of the range of compelling macro issues in many cases creates a 'burning deck' where urgent action is required. Across all sectors, the Blue Economy is a convergence between the commercial desire to capitalize natural resources, and the need to protect the health and diversity of the oceans. Technology has a huge role to play in achieving that balance. However, technological innovation is not always easy.

Established ocean industries can be set in their ways. Decades-old systems are embedded, so the norms, vested interests and practices that underpin them can be hard to shift, especially when they already operate at scale, so large investments of capital may be needed to move towards new ways of working. Emerging industries, conversely, often have less firm grounding on which to build convincing arguments for investment. When science is racing hard to solidify into sound business models, products and services, it can be harder to convince the 'customer' when there isn't a robust track record of delivery as comfort for investment decisions. In contrast to land-based technology development, the seas and oceans can sometimes provide a harsher test bed for innovators as issues related to stability, connectivity, access, safety and the need for robustly ruggedized equipment can all push the research and development bar a little higher.

Specific seaborne issues aside, though, many of the challenges and opportunities as they relate to the use of emerging technology are hardly unique to the Blue Economy, so several times throughout the book we point across to the use of technology in other industries. This helps both to highlight that ocean-focused industries are not alone either in the barriers they face or the shifts in cultural or business practice that need to be factored in to developments, and to point out where some breakthroughs have been made that may offer useful points of learning.

When considering the Blue Economy at its broadest, it is ironic that on the one hand we appear to be plundering and polluting the oceans to unsustainable levels, while on the other we have barely started to explore their socio-economic potential. Perhaps more sensitive and dynamic consideration of the latter could reduce the impact of the former. What is certain, though, is that innovative technology has the potential to be a force multiplier for good intentions in the Blue Economy, as we hope to illustrate in the chapters ahead.

References

CESA (2016) *Global Trends Driving Maritime Innovation*. www.waterborne.eu/media/20004/global-trends-driving-maritime-innovation-brochure-august-2016.pdf (archived at https://perma.cc/4VCJ-S2D8)

OECD (2016) *The Ocean Economy in 2030*. https://read.oecd-ilibrary.org/economics/the-ocean-economy-in-2030_9789264251724-en#page1 (archived at https://perma.cc/FRP6-QZZM)

United Nations Conference on Trade and Development (2018) *Review of Maritime Transport 2018*, UNCTAD.https://unctad.org/en/pages/PublicationWebflyer.aspx?publicationid=2245 (archived at https://perma.cc/4X4V-UMTL)

02

Shipping

Navigating an unprecedented sea of data

Perhaps more than any other sector that will be featured in this book, the shipping industry is the focal point of so many cross-cutting issues and challenges, mainly because of its sheer size and complexity. The shipping industry is a truly global, multidimensional and varied industry that spans all scales from a single pleasure craft, through passenger ferries and tramp steamers to luxury cruise liners (covered in greater detail in Chapter 5), massive super-tankers and the latest queens of the oceans, container ships.

The world's economy trades by sea. Around 80 per cent of global trade by volume, and over 70 per cent by value, is carried by ships and handled by ports worldwide. The United Nations Conference on Trade and Development estimates that the operation of merchant ships contributes about £290 billion in freight rates within the global economy, which is equivalent to roughly 5 per cent of the entirety of world trade (United Nations Conference on Trade and Development, 2018).

There are over 50,000 merchant ships trading internationally, servicing every industry and transporting every kind of cargo to all corners of the globe. The world fleet is registered in over 150 nations and manned by over a million seafarers.

The world's commercial fleet has grown rapidly in the past quarter of a century. At the start of 2017 the global fleet totalled 1.86 billion deadweight tonnage (dwt), compared to 621 million dwt at the start of 1992, exhibiting a growth multiple of three over the period.

World seaborne trade (by tonnage) is dominated by Asia, with the Americas and Europe vying for second place in the global league table. Asian ports also dominate league tables for vessel port calls, capturing 45 per cent of the global total, followed by Europe at 35 per cent and a significant drop to North America at 5 per cent. Developed countries own 60 per cent of seagoing merchant vessels, with developing countries in Asia claiming 36 per cent.

As we see often with Blue Economy predictions, the pivot point of maritime trade is expected to shift from West to East in the coming years, driven by China's rapid economic expansion. The next 15 years will see Southeast Asian nations, the Indian subcontinent and emerging economies of South America drive business growth in this large and complex market.

International shipping companies have been confronted by the challenges of the global recession, compounded by oversupply of vessels ordered during years of commercial comfort and further exacerbated by long lead times for vessel build and disposal.

The shipping market's reaction in the years following the recession of 2008 demonstrated its flexibility and adaptability. Taking advantage of corresponding low fuel costs, low vessel values and oversupply, the industry has adopted low-speed steaming and floating storage as effective mechanisms to address the short- to mid-term market conditions. Alliances and takeovers have become more prevalent.

Despite exhibiting such flexibility and resilience, the headlines in October 2018 associated with the launch of a new report featuring the views of senior maritime stakeholders painted a picture with noticeably dark skies. 'Senior maritime stakeholders deem industry not prepared to deal with global issues,' said one (Chambers, 2018). The Global Maritime Forum (2018) collated the thoughts of senior maritime stakeholders from more than 50 countries and presented them to the world in the report *Global Maritime Issues Monitor 2018*. This highly accessible report focused in on 17 major issues that could possibly have an impact on the sector in the following ten years. The participants ranked these by their likelihood of happening and then scores were ascribed to each according to how prepared the maritime leaders felt they were to deal with them. The issues included: cyber attacks and data theft (deemed most likely to happen, with a rating of 3.67 out of 5); air pollution; workforce and skill shortages; terrorism; natural environmental disasters; pervasive corruption; and increased piracy.

The issue that the senior respondents felt the least prepared for was cyber attacks and data theft. This should perhaps come as little surprise. As evidenced throughout this book, the Blue Economy is becoming ever more reliant on digital infrastructure and high-powered services driven by data. The potential for such operations and services to be disrupted by cyber attacks will therefore only increase in the coming years. At a time when maritime companies have already been squeezing their already tight profit margins to invest in innovative digital systems, they then may discover that a whole new level of risk also needs to be managed, requiring additional resource.

The industry's major wake-up call in this area came in the summer of 2017 when global container shipping company A.P. Moeller Maersk estimated that a cyber attack it had endured could cost it $300 million in lost revenue. The attack was part of the infamous NotPetya ransomware incident, where hackers infiltrated and locked down entire networks. Each individual system user was faced with the stark message that they could only regain access to their terminal if they pay $300 in universal digital currency bitcoin. The original target for the attack was Ukraine, where hospitals, power companies, airports, banks, ATMs and card payment systems were all seriously affected, but the ransomware soon spread further. It rapidly swept through the Maersk system on 27 June, closing computer after computer before the central IT team could react, rendering most office employees entirely unable to do any work in this digitally reliant world. For two days, Maersk Line was reportedly unable to take bookings from customers, though they stressed that no third-party data was lost as a result of the attack.

In October 2017, three months after the incident itself, global maritime outfit BW Group announced that their cyber defences had also been breached over the summer, hard on the heels of the Maersk attack. While the company did not reveal the full nature of the attack (though they did say that it was not ransomware), BW Group insisted that the issue had been fixed and that they managed to conduct 'business as usual' during the unauthorized intrusion into their systems. No estimate was provided as to the potential costs of the attack.

More recently, the Port of San Diego reported in September 2018 that it had fought off a ransomware cyber attack on its systems. Again, a system takeover was attempted, with the remote perpetrators demanding payment in bitcoin to rectify the situation. As ports are responsible for coastal security matters as well as commerce, this incident was investigated by the US Department of Homeland Security and the Federal Bureau of Investigation (Porttechnology.org, 2018). The San Diego incident occurred a week after a similar incident had affected the Port of Barcelona, though they released fewer details.

Even these appear to be far from isolated incidents. As in every industry, the maritime sector is under cyber threat like never before. In London in November 2017, ship owners and oil majors at the annual Tanker Shipping and Trade Conference candidly released statistics that highlighted the constant barrage of threats their organizations routinely face. BP's Chief Financial Officer Guy Mason revealed (Bergman, 2017) that in the previous year his company had managed to intercept 614 million phishing emails that attempted to obtain sensitive information such as user names, passwords and financial details. That equates to 1.7 million such emails a day, 70,000 per hour, 1,200 a minute – a frightening and never-ending barrage of attempted fraud. The size of the threat was then corroborated by Paul Rodgers, the Chief Executive Officer of Euronav, the largest New York Stock Exchange-listed independent crude oil tanker company in the world, who said that his organization endured 50,000 daily attempts to breach its cyber security systems.

Both BP and Euronav should be applauded for their honesty. As companies work to counter ever-more sophisticated cyber attacks, this kind of transparency can only be positive so the entire industry can understand the size of the problem and work together to defeat it. While email phishing could now be a relatively longstanding type of threat, new maritime security challenges are emerging all the time.

Playing 'spoof'

A novel factor to worry about came to prominence in 2017, relating to ships' global positioning system (GPS). We discuss in Chapter 6 why some nefarious seafarers may wish to turn off or otherwise mask their digital footprint to hide their positioning and avoid others' knowledge of or interest in their activities. This, however, was different, as it appeared to be happening on a larger scale – and it was not being driven by the captains or ship owners themselves.

The issue came to light when the US Coast Guard Navigation Center received a worrying report (The Maritime Executive, 2017). A concerned captain in the Black Sea sent the Navigation Center pictures of his systems erroneously showing him that he was 25 nautical miles away from his actual location. The issue was not exactly difficult to spot, as the GPS equipment was attempting to tell the captain that he was at a land-based location near to Gelendzhik airport in Russia.

A simple case of accidental interference or faulty equipment? Not quite. The US Coast Guard immediately confirmed that no exercises were being conducted at that time so the GPS should be accurate to three metres. It also eventually became known that another 20 vessels in the locale were receiving the same confusing information. As they would have used different GPS equipment, how could they all be witnessing the same anomalies?

A subsequent investigation was undertaken by the Resilient Navigation and Timing Foundation, a non-profit organization based in Alexandria, Virginia, USA, that advocates for better security related to the global positioning system. They worked with maritime data and analytics company Windward Ltd to explore the issue in greater depth, eventually diagnosing that the incident related to the intentional 'spoofing' of the GPS signal, ie purposefully interfering with it to cause it to provide incorrect time or location information. It was also not an isolated incident, as two other cases of similar mass interference were identified through the use of targeted algorithms. This presents a severe threat to a system that is essential to seafarers of all types.

Further research in April 2019 by the Centre for Advanced Defense (Edwards, 2019) underlined the emergence of such disruptions and laid the blame for such activity squarely at the door of the Russian Government. The study suggested that 1,311 civilian ships had been affected in 10 locations over the study period, with 9,883 individual incidents reported or detected, the majority of which took place in the Crimea, the Black Sea, Syria and Russia. Whereas previously the Russian State had been believed to use such techniques to help to disguise the whereabouts of President Vladimir Putin, the report suggested that this practice may have been extended to take place near to protected facilities on the Black Sea coast. The other worrying factor is that the technology to undertake such spoofing activities is relatively cheap and readily available on the open market, so this is far from simply a complex military issue. When we witness how much disruption can be caused with the launch of a simple drone near to an airport, the potential to disrupt shipping lanes and other seaborne activity is obvious, and can only be a major concern.

Both the UK Government's Blackett Report in 2017 (Government Office for Science, 2017) and a subsequent report by specialist policy and economics consultancy London Economics (Sadlier et al, 2017) highlighted the potentially severe disruptive risks associated with Global Navigation Satellite System (GNSS), and the operations that depend on it. For example, the report estimated that an attack that disabled GNSS in Britain would cost about £1 billion every day the system was down. In the maritime space these vulnerabilities are already significant. Factors such as space weather can cause ships to appear on charts at inaccurate locations, but intentional jamming and spoofing is a whole new ball game.

As a side note, NLA International Ltd is pleased to be playing an active role in exploring some of the problems that relate to these issues. The Maritime Resilience and Integrity of Navigation (MarRINav) project is exploring vulnerabilities in position, navigation and timing (PNT) solutions. Funded by the European Space Agency's NAVISP programme, MarRINav is considering these risks and exploring how other PNT solutions can complement GNSS to provide all mariners with greater resilience and confidence in what is now recognized as critical national infrastructure.

The NLAI-led consortium includes the University of Nottingham, the General Lighthouse Authorities Research and Development team, Terrafix, London Economics, Innovate UK, Taylor Airey and University College London. Together we are exploring how a system-of-systems solution might emerge primarily for maritime PNT resilience and integrity while recognizing that findings may be applied to other sectors such as transport and logistics, emergency response, security, financial services, power distribution and telecommunications. Protecting GPS from spoofing, or coming up with workaround systems, is essential to the future of marine navigation.

Rising to the cyber challenge

In the face of such cyber threats – new and old – many more in the industry are beginning to take on leadership roles. While some attempt (unfairly, in our view) to portray the shipping sector as being technologically backward and slow to respond to challenges, it's clear that the industry's increasing reliance on digital systems (both onshore and at sea) data means that avoiding urgent action in relation to cyber security issues is simply not an option. Thankfully, there is plenty of encouraging news to show how the industry is strengthening its resolve in the face of this growing threat.

In October 2018, Finnish smart technology company Wärtsilä announced the launch of a new International Maritime Cyber Centre of Excellence in Singapore (World Maritime News, 2018). Hailed as the first of its kind in the world, it promises to catalyse thinking, awareness and action in the face of this significant and growing threat to the entire sector.

In addition, KPMG and Kongsberg announced a partnership to offer specific maritime sector-focused cyber security solutions (Martin, 2018), and we can expect many more such announcements in the next couple of years, as in June 2017 the International Maritime Organization (IMO) added an important regulatory driver into the mix. They announced that ship owners and managers had a deadline of 2021 to provide evidence that they have incorporated cyber risk management into their ship safety processes. To underline the importance attached to this endeavour, companies who do not comply may risk having their ships detained. Whatever the pros and cons of the IMO decision, we can certainly only expect the maritime cyber security market to grow as ever-more determined and sophisticated attacks are launched. While the timescale has been criticized by ship owners, such tight regulations quite often lead to a flurry of new technology options rushing to the market.

Nation states are also stepping up to meet the maritime cyber challenge. In January 2019, for example, the Danish Ministry of Industry, Business and Financial Affairs launched a range of new initiatives within a new cyber strategy for its shipping industry (Hellenicshippingnews.com, 2018). Planned activities included establishing a new Danish Maritime Cyber Security Unit; more effective awareness raising, collaboration and knowledge sharing initiatives; and establishing more coherent and collaborative contingency and warning plans in case of serious cyber security incidents in the Danish maritime sector.

Tackling the energy challenge

Referring back to the *Global Maritime Issues Monitor*, while cyber security and the potential for data theft were singled out as the issues that the senior maritime respondents felt the least prepared for, energy efficiency was predicted by them as the area likely to have the greatest impact on the sector in the coming decade. Respondents suggested that improving the energy efficiency of vessels and deciding on a strategy to reduce greenhouse gas emissions are two very important first steps on the path towards this goal.

The role of technology is crucial to addressing these needs, and maritime executives will be faced with an ever-expanding menu of tech options in the coming years, as again regulatory drivers are entering the fray to compel businesses to act. The range of energy efficiency technologies within shipping is already broad and expanding still further as new technology options progress from the science base into commercialization. These options are often grouped under four headings: ship design, propulsion (see the following section), machinery, and operational strategies/modifications.

Ship design contains practices as diverse as the optimization of hull dimensions and/or the aerodynamics of the superstructure; optimization of the hull and propeller interface; optimization of the ballast and trim; and use of newly available lightweight construction materials.

Within the machinery section, efforts can focus on engine adjustments and tuning; diesel–electric drives; and reducing onboard power demand through advanced power management systems or complementing it by utilizing solar power.

Operational strategies and modifications encompass applying specialist coatings to the hull; utilizing engine lubricants and applying scrubbers (equipment fitted in the exhaust system of a ship that absorbs carbon dioxide); and voyage optimization/weather routing. This last option has been a concern of route planners and captains for centuries, but like many practices featured in this book it has received something of a makeover with the introduction of new Big Data capabilities.

We look in Chapter 5 at how sensors, digital connectivity and machine learning are being put to good use enhancing the customer experience in the cruise industry. However, such investment in technology and data capabilities is not only reserved for the highest end of luxury passenger services. Whether they are aware of it or not, passengers on what may be a more mundane segment of the shipping sector – ferry services – are also beginning to benefit, or will soon, from improvements driven by the possibilities of advanced data analytics.

Dr Lars Carlsson completed his PhD in Naval Architecture and Scientific Computing at Gothenburg's Chalmers University of Technology in 2003 and was subsequently employed in a variety of roles at pharmaceutical firm AstraZeneca that focused on computational and machine learning. In May 2018, however, he returned to his navigational roots when joining Swedish ferry operator Stena as their new Head of Artificial Intelligence, a world first for any ferry company. The appointment of a head of artificial intelligence was an important stage in the company's public drive towards becoming an 'artificial intelligence (AI)-first' organization. In the past few decades, many companies from across the industrial spectrum have made similar announcements to highlight how they plan to be operationally and strategically driven by new technological possibilities – from 'mobile-first' when the smartphone began its inexorable consumer explosion, to 'data-first' when new analytics capabilities were able to extract value in new ways from the 'four Vs' of Big Data (volume, variety, velocity and veracity), heralding a new dawn of business intelligence.

As so often in the digital space, tech behemoth Google were one of the earlier companies to nail their colours to the mast of the next wave of innovation. Back in 2016, Google Chief Executive Officer Sundar Pichai first signalled the company's intended progression from being a 'mobile-first' company to an 'AI-first' organization (D'Onfro, 2016).

Industrial AI has been in development for many years, arguably since the invention of the programmable computer in the 1940s. For decades, the term most often collided with public consciousness when a computer was pitted against a chess master or as the focal point of a world inevitably gone wrong in a science fiction novel or movie. However, it wasn't until the progression of new data processing, deep learning and machine learning capabilities in the past two decades that its true potential began to be understood and exploited.

For Google, far from just providing search results on a mobile phone, this meant moving towards providing intelligent assistance to help users with their needs – *in context* – and especially, still, on mobile devices. In product terms this means, for example, Google Photos applying algorithms to automatically identify people, places and objects within your photos, to help you organize your content. Traditional search functionality instantly becomes more powerful when geo-tagged information is folded into the mix, both where you are now and based on your previous travel activities and preferences. The argument goes that the more such layers of information become automated, rather than having to be spelled out by the user, the more seamless and useful the service becomes.

And this brings us back to maritime vessel efficiency. In September 2018 Stena's Lars Carlsson revealed that the company was launching a pilot project, in partnership with Hitachi, to put an AI system to work to suggest the most fuel-efficient route for its ferries to follow (SAFETY4SEA, 2018a). The AI system utilizes all available data to model efficiency on a number of potential routes, before finally proposing the most optimal for fuel optimization.

Ultimately, this AI system is being adopted by Stena to help them meet their stated annual fuel reduction target of 2.5 per cent. According to industry body InterFerry, the ferry industry is similar in operational size to the commercial airline industry, transporting as it does approximately 2.1 billion passengers per year, plus 250 million vehicles and 32 million trailers (note that these figures do not include China). So, if similar AI-driven gains could be made across the entire industry, the economic and environmental benefits would be hugely welcomed.

Plain sailing?

Addressing another of the four energy efficiency themes, innovative propulsion measures also focus on new build and retrofit options, in this case for thrusters, propellers and rudders, but also feature more innovative options such as electric propulsion and wind power.

We cover the use of innovative sail technology within the cruise industry in Chapter 5, but this innovation is also being used within the broader shipping sector. In September 2018 global aerospace manufacturer Airbus announced that it was experimenting with wind power to help reduce emissions and cut costs in moving cargo at sea (Wall, 2018). The firm uses large cargo ships to move aircraft parts around the world and announced that it would be introducing innovative sail technology on one of its three roll-on, roll-off vessels. The company revealed that it hopes that the 5,382 square foot sail will reduce annual fuel spend by approximately £865,000 and – crucially – reduce carbon dioxide emissions by about 8,000 metric tons a year. This joint benefit – providing a direct commercial return on investment at the same time as providing significant environmental gains – bodes well for further adoption across the industry.

Maersk Tankers is also engaging with this new technology. They announced their initial investment in sail technology a few days before Airbus, and revealed that they have ambitions to cut 7 to 10 per cent of a ship's fuel use when certain conditions are in play. So hopeful are they of the initiative's potential that a Maersk spokesman has suggested that the use of wind-harnessing technologies could see wind-assisted seaborne companies altering their trading patterns to maximize the wind energy available (Informare.it, 2018). If a trading route is longer but takes the vessel on a course that provides far more favourable wind support, the cost of fuel saved may well be significant enough to warrant the extra time at sea. In times when fuel costs are volatile, the attraction of this option is tangible. With Maersk and others hoping eventually for double-digit percentage savings, the results of these emerging initiatives will be worth following.

From reaction to prediction

Another exciting advanced sensor- and data-driven approach to vessel and fuel efficiency (and, as we shall see, much more) is one that has a long backstory and has been progressing steadily in recent years. It's one that is beginning to show great promise and attract serious investment. Its origins lie in the aviation industry. In the 1960s it was discovered that a large percentage of failures on complex aircraft were not age-related (eg due to fatigued components) but more random. This therefore meant that such failures were not adequately addressed by preventive (scheduled) maintenance. Further research of these issues led to the development of a new maintenance strategy called reliability-centred maintenance (RCM).

RCM is a process for determining a maintenance strategy that considers safety, operational benefit (the availability of systems for use) and economy. RCM advocates the implementation of a *predictive* maintenance scheme in which the condition of in-service equipment is monitored in order to *predict* when maintenance should be performed, rather than just to expect it to fall into a regular routine of operational intervention. The aviation industry embraced this new approach with open arms, with take-up increasing several fold within the first five years of its availability (1964–69).

Maintenance has therefore evolved from a reactive process, performed after a functional failure, to a preventive activity where items are overhauled or discarded according to a time schedule. Preventive maintenance assumes that a component has a defined lifetime, after which its failure rate increases. However, estimates of lifetime often have large uncertainties. Hence, scheduled maintenance is often performed too early or too late, resulting either in high costs due to unnecessary replacements or functional failures respectively. Even worse, a component that cannot be inspected from the outside is often disassembled and inspected on schedule, with the risk of introducing faults during inspection or re-assembly, thereby leading to failure shortly afterwards. All of this is costly and can be wasteful.

Despite such a promising case study to learn from in the aviation industry, the progress of condition-based monitoring within shipping has been variable. The monitoring mostly takes place at the component level and the sophistication is dependent on the component maker. Standard sensors measuring quantities such as temperature, pressure, vibration and strain are normally used, but research into new sensor types for detection of failure modes is also being carried out.

In the maritime environment, main engines are normally the components with the most sensors and sophisticated monitoring systems, due to the risks associated with main engine failures. Damage statistics show that the main engine is the single component responsible for most serious damage events on ships. Condition-based monitoring has also been expanded to other components in the propulsion system, such as gears, bearings and propellers. In addition, critical auxiliary components and systems are subject to sensor-based monitoring.

Put simply, the multifaceted benefits of condition-based monitoring could be significant. For example, using an effective condition-based monitoring system to enable the timely and effective in situ repair – either on board or ashore – of, say, a crankshaft or its pins could prevent the loss of the entire engine, and also guards against potentially must costlier incidents such as blocking ports or strandings at sea.

Two per cent of the global fleet of ships currently use condition monitoring systems (some 2,150 ships). While it is not possible to draw quite such a straight line, if the shipping sector is to follow the trend seen in the aviation industry, this may increase to create a market of, at a rough approximation, 40,000 ships over the next five years. And commercial providers are doing their best both to invigorate and attempt to meet that need. A consistent theme is emerging across their endeavours.

Finnish ship operator Bore has revealed that it had utilized the ClassNK-NAPA GREEN condition monitoring system to unlock fuel savings of between 4 to 6 per cent across a fleet of three roll-on, roll-off ships: the 2,863 lane metre (lm) *Bore Sea*, 2,863lm *Bore Song* and 1,606lm *Seagard*.

Most SkySails performance managers have been installed on tankers, bulkers, multi-purpose vessels and containerships. So far, according to the company, customers of this product have seen fuel savings in the range of 3 to 10 per cent.

The breadth of data gathering that can drive such systems is best exemplified by statistics released by James Fisher Mimic. They state that their Turbocharger monitoring technology is already used on over 150 turbines throughout the cruise industry, with tankers and bulkers their secondary market. According to the company, installing their system on P&O Cruise's *MV Britannia* involved establishing an infrastructure that would pull in data from over 200 ship-based assets that included air conditioning compressors, boiler feed pumps, ballast pumps, fresh water pumps, chilled water pumps, salt water pumps, fuel centrifuges, heat exchangers, steering gear and fuel pumps.

James Fisher Mimic has also highlighted the potential for an impressive return on investment on their products, explaining how installing a £15,000 condition monitoring system could save a company 10 times that amount. Speaking at a conference, Engineering Manager Martin Briddon explained how a lube oil pump failure could cost £57,000 and take three days to repair, leading to off-hire losses of £90,000 and other costs, such as management time and increased insurance premiums (Wingrove, 2015). Speaking at the same conference, French energy company Technip (which has subsequently merged with FMC Technologies to form TechnipFMC) reported that it was saving approximately £770,000 per year on equipment replacement and downtime costs as a result of introducing condition-based monitoring (CBM) on vessels (Foxwell, 2015), led by a pilot project on subsea construction vessel *Apache II*. Solving thruster issues, improving hydraulic systems and a reduction in vessel downtime all contributed to this headline impact.

Some operators move towards implementing CBM due to major outages of service that they come to realize could have been prevented. Offshore engineering group McDermott is one such company; it introduced CBM across its fleet after it experienced a heavy £1.5 million loss due to a gearbox failure that compelled it to shut down one of its vessels for more than 30 days. Investigations revealed that a fully functioning CBM could have prevented the failure and the loss, so they went ahead with that change of approach.

In the past few years, the authors have spoken extensively to ship owners and other maritime leaders about the emerging potential of condition-based monitoring services – some who have already started to buy and implement systems, and others who are still weighing them up as potential options within their suite of fleet management systems. It is this element that appears to provide the greatest value to ship owners and operators.

While CBM systems have been introduced as having a major impact on fuel efficiency, many operators we have spoken to are far more interested in their ability to reduce the amount of 'down time' they experience – where ships are forced out of action due to the failure of systems or individual components. As one president of a ship management company told us: 'The most important thing is that all equipment is available at all times.' Reducing down time may currently be a more important and valuable offer within the CBM market than fuel efficiency. Put baldly, if a vessel isn't available it can't earn money. Reliability and predictability are king. Many fleet operators carry a large range of spare parts at all times to minimize potential downtime. The option to not have to do so, as they have greater confidence in when repairs will be needed, will also help operational efficiency and cash flow, as the need for large stocks of spare parts will be minimized.

With all these benefits in mind, though, the adoption rate of CBM systems within the shipping industry is still very low. The question must be asked as to why more companies are not taking up the offer. The answer is complex, and multi-faceted. First of all, while potential returns on investment for such systems are beginning to be acknowledged and understood, finding the investment to purchase systems is often still not easy. Many shipping companies are owned by investment funds and banks, who tend to seek results on a quarter-by-quarter basis, so talking of returns on investment in terms of years is often not enough to unlock the hard cash needed to purchase such systems.

Second, as explained in other chapters, owners and operators are approached all the time with new digital solutions promising to revolutionize their efficiency and profitability. As well as holding their own on the value proposition, then, CBM solutions also need to be developed so that they can integrate seamlessly with other systems (new and emerging) to avoid a messy proliferation of standalone software.

Beyond that, CBM requires a significant shift in thinking. It doesn't simply provide a new and improved component that is introduced when something fails; instead, it requires all staff and management to buy into a new way of doing things. Current diagnostics can be dependent on the engineer and/or ship's skipper having a feeling that things are or are not 'right' with an engine. While such staff may be enlightened enough to see CBM systems as enhancing their ability to operate effectively by making the right decision at the right time, there is enough 'noise' around all industries about machines making human endeavours redundant that it is understandable that some may see them as a threat. Change management is difficult in any industry, and no less so in shipping, so providers of condition-based monitoring systems still find themselves in essence creating a new market, relying on completely new ways of thinking and changes to long-established norms, rather than simply trying to slip new tools into existing operational models.

Looking more broadly across a fleet, larger issues also emerge. Additional infrastructure costs may be needed to extract the maximum value from such data-informed possibilities. One ship's engine alone can produce up to one terabyte of data a month. With the world's largest shipping line owning 700 ships, storing all the engine data may be expensive. Ship owners may worry that it is not yet known which elements of the processed data need to be stored. And how long does such data need to be kept, enabling the kind of deep machine learning across multiple engine data sets that will really provide value in the era of high-performance computing?

The need for predictive condition monitoring becomes more important in relation to one of the emerging technologies that cuts across almost every Blue Economy sector – autonomous vessels. How can you monitor the condition of systems when nobody is on the ship? While we have featured autonomous and remotely operated vessels in various sections of this book – as floating platforms capable of carrying a broad range of sensors that add value in domains as diverse as oceanography and maritime security – their promise within the shipping industry is also being explored with some vigour. Rolls-Royce can be seen to be leading the way in this respect. In December 2018, in partnership with Finferries, the State-owned Finnish ferry operator, the company demonstrated a fully autonomous ferry in the archipelago south of the city of Turku (Jiang, 2018a). In front of 80 assembled guests, the system successfully demonstrated its autonomous operations and collision avoidance capabilities – driven by sensors and artificial intelligence rather than human intervention.

While autonomous ferries may stay close to shore, it is worth noting the Rolls-Royce has announced that it expects to have its own autonomous deep sea cargo ships in operation by the year 2030 (Jiang, 2018b). If such vessels are ever to be fully autonomous, they obviously cannot rely on the nous and intervention capabilities of savvy seafarers, so condition-based monitoring capabilities will become ever more important and valuable. And in case there is any doubt that autonomous shipping over large distances is possible, the first Atlantic crossing by an autonomous vessel was completed in September 2018 (Belfasttelegraph.co.uk, 2018). Said to have successfully tackled every wind state from calm to a strong gale during the crossing, the *SB Met* Sailbuoy took 80 days to travel the 3,000km from Newfoundland in Canada to Ireland.

The power of sharing

There is an additional benefit to the implementation of a more proactive equipment monitoring and maintenance system. Safety is undoubtedly the most critical key performance indicator in any Blue Economy domain, and that is no different in shipping. Statistics from the IMO show that machinery failure causes nearly 25 per cent of all marine accidents. Furthermore, most machinery failures take place immediately after a maintenance activity. While it is not possible to draw an entirely straight line about this, the case could be made that, naturally, the additional pressure to get unexpectedly out-of-service systems or ships patched up and back into operation could have an impact. Therefore, being able to plan ongoing or shore-based maintenance in and around planned customer activity can only ease such pressures, potentially allowing for greater periods of testing post-maintenance.

It is worth noting that it is not only the monitoring of mechanical systems that can lead to enhanced crew safety. In February 2018, Shell's vice president of shipping launched HiLo, a predictive modelling tool for accident prevention in shipping (SAFETY4SEA, 2018b). Noting the alarming statistic that shipping has a fatal accident rate five times that of construction, the launch materials outlined that the system aims to prevent 'high-impact' events like explosions, collisions and groundings by recording the frequent low-level incidents that are precursors to major incidents. With HiLo, data from smaller, on first inspection non-threatening, incidents can be interrogated and analysis can then be applied to target the specific areas in which safety should be improved onboard vessels. Rather than being inconsequential, these smaller incidents can all be viewed collectively and – with the right data analytics tools – come together to create an early warning system.

Shell is joined by Maersk (and a much broader collection of shipping companies), Lloyd's Register and the Lloyd's Register Foundation in this initiative. What is perhaps most noteworthy about this project is the news that all of the shipping companies involved have contributed incident data to the system in its trial phase. Access to data is the key to the success of any such initiative, so it's hugely encouraging to see so many companies understanding that viewing their own data in isolation will only ever be a fraction as powerful as seeing it supporting much broader analysis. When everyone shares (within agreed sensible boundaries), everyone wins.

Conclusion

When considering the digital and connectivity agenda in shipping, it is important not only to think of products, services and the growing promise of the Internet of Things agenda that allows systems of machines to connect with each other and share information. There are also very human gains to be made in the digitally connected world of shipping. Being able to resolve crew needs was one the original primary driver of satellite connectivity for shipping, so that owners were able to facilitate regular contact between home and family while crew members were deployed for long periods. Many studies over the years have exposed the potential negative effects of isolation on those working at sea for long periods of time, so the truly connected agenda in shipping is a huge leap forward on that basic level.

Having been pushed to innovate by that human need, though, owners and operators are now progressing the benefits of additional connectivity, what operational efficiencies could result from better digital business tools, more efficient information exchange, shared data sets and so on. In short, the potential that various forms of digitalization offer to the shipping sector needs to sit at the very heart of future plans for the industry, in all its parts.

Shipping markets are still recovering from global financial uncertainty, profit margins are as tight as ever and excess tonnage prevails, so companies can be reluctant to invest. The challenge is therefore to prove that digitalization will streamline operations, reduce overheads and improve margins – technology should be a way to save money, not spend it.

As we have discussed in this chapter, specifically in relation to the relatively slow take-up of condition-based monitoring systems in the shipping industry, the barriers to proper digitalization in the sector are many, varied, complex and in some cases stubborn. In particular, it can be difficult to envision or drive digital progress without adopting a holistic view. A proper understanding is required at board level that bold measures, not short-term sticking plasters, are truly required to embrace digital opportunities to the fullest. Thankfully, some leaders are very much beginning to embrace this mindset and take appropriately far-reaching steps. For example, in January 2019 the president and chief executive officer of Japanese shipping firm Kawasaki Kisen Kaisha (K Line – Japan's third largest ship owner) announced a major corporate restructuring whose main purpose was to promote the take-up of digitalization throughout the company (Chambers, 2019). Specific reference in the statement was made to the possibilities of artificial intelligence and the Internet of Things. If a major firm currently celebrating the centenary of its inception can adopt this approach, so can others of more varied longevity.

The authors are pleased to participate regularly in one of the international fora that act as the fulcrum for emerging ideas in this domain. The Digital Ship Chief Information Officer conference series provides a very useful platform for regular engagement in this exciting area, and a progressive space in which to explore in detail the pros and cons of digitalization in shipping with diverse groups of stakeholders. Regular NLAI engagement in these fora over the past few years has provided an opportunity to chart progress in a range of related areas that include satellite communications and connectivity, the potential of blockchain technology for shipping, Big Data and IoT, and cyber security. Each of these has its challenges and opportunities, but overall we are left with the sense of an ever-maturing, steadily consolidating sector view that digitalization may well be the game-changing solution for improving the efficiency of shipping operations, their interface with ports and their contribution to the huge and complex global logistics network.

The application of AI within the shipping industry and the promotion of data-driven, condition-based monitoring are two interesting examples of the profitable use of new information in the digital age. Greater understanding of the potential of (and, at the very basest of levels, the need to collect) data needs to pick up pace in shipping industry. Though it is far too early to tell, the greatest gains may well be found when large volumes of data – that on first review seem unrelated – are brought to bear to improve specific services.

Case studies in other domains repeatedly highlight that the answers to problems may lie in forgotten or uncorrelated data. Perhaps the most celebrated case here is where New York City authorities used predictive modelling in 2011 to pinpoint buildings with a higher than normal fire risk. Whereas previously they would only have looked at standard data such as building age and historical fire risk, the new data-driven model attempted to pull in a much broader wealth of base information. Analysts focused on factors like missed tax or utility payments – which suggests neglect – and nearby crime and accident rates. Then, by visiting only those buildings that the data pointed to, fire inspectors were able to evidence that their model enabled them to find many more risks than if they'd simply gone door-to-door (Heaton, 2015). The model has continued to develop in subsequent years and shows how progress within the digital shipping environment may push the industry towards all manner of new opportunities.

Within this domain, connectivity is of course simultaneously potentially an enabler and a significant blocker, as we see in so many Blue Economy sectors pushing towards varying forms of digitalization. Having spoken regularly with ship owners, though, the conversation on connectivity at sea has evolved over the past five years from concern about the cost and adequacy of satellite communications to a growing realization that satellite service providers have actually begun to understand end user needs, modified their systems and hardware accordingly, and reviewed products and services in such a way that affordable, ubiquitous 1mb data connectivity services within the maritime sector are a realistic prospect in the near term. This will then hopefully provide the lifeblood for a much broader range of digital services to flourish. Admittedly, such connectivity is not yet the equivalent of terrestrial services (and terrestrial will for obvious reasons be difficult to match for the foreseeable future) but it's undoubtedly a vast improvement on past performance, clearly paving the way to digitalization in all aspects of shipping operations.

What will help to underpin ongoing progress will again need the human touch – the openness on display at events like the Digital Ship CIO Forum, starkly honest research such as that contained in the candid *Global Maritime Issues Monitor* and the willingness of individual company leaders to be open about the size and complexity of the cyber security challenges they are facing.

The development and adoption of condition-based monitoring systems in the shipping industry highlights the difficulties that technology innovators face in many related Blue Economy domains. Even though many companies can show concrete examples of tangible returns on investment related to their systems, the adoption rate across the industry is still surprisingly small. Difficult investment conditions, the desire for very short-term savings, and the need for staff buy-in within what could be a complex change management process all combine to make it a difficult 'sell' to ship owners, despite the potential gains. Those who are ultimately able to tackle vessel efficiency and condition-based monitoring strategically and holistically – integrating and supporting as many additional and complementary systems at the same time – will make the most gains. But that is a tough prospect for those who have been developing one system, no matter how powerful, and need to commercialize before their own investment pot runs dry and they go to the wall.

Finally, as all these approaches push the shipping sector further towards reliance on digitalization and the seamless availability of data, the reverse side of that coin could well be viewed as the shipping industry being more vulnerable to cyber attack. Speaking regularly to chief information officers at associated conferences, there is never a sense of satisfaction or schadenfreude when a competitor is revealed to have suffered a cyber attack; the reaction is always far more closely aligned to one of 'There but for the grace of God go I.' It is therefore encouraging to see that this is reflected tangibly in meaningful industry exploration of and collaboration against the new levels of cyber threat in the maritime industry. For example, this was underlined in December 2018 when some of the world's largest shipping associations combined to update their shared cyber security guidelines (Bergman, 2018). The Baltic and International Maritime Council, the International Association of Independent Tanker Owners, the Oil Companies International Marine Forum and the World Shipping Council all came together to collaborate on the project, which provided sensible guidance for the development of strategies and underpinning operational processes, as well as offering anonymised examples of real-world cyber attacks on ship owners and operators, rather than just hypothetical scenarios.

In what is becoming an increasingly critical area, where standardization of terminology, tools and approaches can only aid understanding and subsequent implementation, such collaborative endeavours are to be hugely welcomed, and bode well for the next stages of the fight against cyber crime at sea.

References

Belfasttelegraph.co.uk (2018) First unmanned sailboat to cross Atlantic celebrates success. www.belfasttelegraph.co.uk/news/republic-of-ireland/first-unmanned-sailboat-to-cross-atlantic-celebrates-success-37281791.html (archived at https://perma.cc/U7XA-9WHL)

Bergman, J (2017) BP and Euronav reveal cyber attack statistics, discuss hijacking threat. www.marinemec.com/news/view,bp-and-euronav-reveal-cyber-attack-statistics-discuss-hijacking-threat_49956.htm (archived at https://perma.cc/LJS5-QKRN)

Bergman, J (2018) Shipping groups publish cyber security guidelines update. www.marinemec.com/news/view,shipping-groups-publish-cyber-security-guidelines-update_56155.htm (archived at https://perma.cc/3JW5-ZSFL)

Chambers, S (2018) Senior maritime stakeholders deem industry not prepared to deal with global issues. https://splash247.com/senior-maritime-stakeholders-deem-industry-not-prepared-to-deal-with-global-issues/ (archived at https://perma.cc/C7WW-JJKE)

Chambers, S (2019) K Line makes sweeping digital reforms. https://splash247.com/k-line-makes-sweeping-digital-reforms/ (archived at https://perma.cc/6H3H-8T4N)

D'Onfro, J (2016) Google's CEO is looking to the next big thing beyond smartphones. www.businessinsider.com/sundar-pichai-ai-first-world-2016-4/?IR=T (archived at https://perma.cc/JN5G-LVNP)

Edwards, J (2019) The Russians are screwing with the GPS system to send bogus navigation data to thousands of ships. www.businessinsider.my/gnss-hacking-spoofing-jamming-russians-screwing-with-gps-2019-4/?r=US&IR=T (archived at https://perma.cc/T9SY-SJ34)

Foxwell, D (2015) Offshore sector making better use of CBM than some. www.osjonline.com/news/view,offshore-sector-making-better-use-of-cbm-than-some_41210.htm (archived at https://perma.cc/2ZMA-RSWK)

Global Maritime Forum (2018) *Global Maritime Issues Monitor 2018*. www.globalmaritimeforum.org/content/2018/10/Global-Maritime-Issues-Monitor-2018.pdf (archived at https://perma.cc/5DHJ-8JDV)

Government Office for Science (2017) *Satellite-derived Time and Position: A study of critical dependencies*. https://assets.publishing.service.gov.uk/government/uploads/system/uploads/attachment_data/file/676675/satellite-derived-time-and-position-blackett-review.pdf (archived at https://perma.cc/79EG-3WRN)

Heaton, B (2015) New York City fights fire with data. www.govtech.com/public-safety/New-York-City-Fights-Fire-with-Data.html (archived at https://perma.cc/65WB-9BCC)

Hellenicshippingnews.com (2018) New strategy for cyber security in the Danish maritime sector. www.hellenicshippingnews.com/new-strategy-for-cyber-security-in-the-danish-maritime-sector/ (archived at https://perma.cc/9DP3-8ZRX)

Informare.it (2018) Harnessing wind power could change trading patterns says Maersk Tankers. https://fairplay.ihs.com/ship-construction/article/4306791/harnessing-wind-power-could-change-trading-patterns-says-maersk-tankers (archived at https://perma.cc/GL8H-9DZN)

Jiang, J (2018a) Rolls-Royce demonstrates fully autonomous ferry. https://splash247.com/rolls-royce-demonstrates-fully-autonomous-ferry/ (archived at https://perma.cc/XAM3-PCRK)

Jiang, J (2018b) Rolls-Royce eyes deepsea autonomous shipping before 2030. https://splash247.com/rolls-royce-eyes-deepsea-autonomous-shipping-before-2030/ (archived at https://perma.cc/Y74C-H723)

Martin, E (2018) Kongsberg and KMPG partner up on cyber security. www.marinemec.com/news/view,kongsberg-and-kpmg-partner-up-on-cyber-security_54489 (archived at https://perma.cc/JDD9-W9UA)

Porttechnology.org. (2018) San Diego suffers cyber attack. www.porttechnology.org/news/san_diego_suffers_cyber_attack (archived at https://perma.cc/K7WV-F3A6)

Sadlier, G, Flytkjær, R, Sabri, F and Herr, D (2017) *The Economic Impact on the UK of a Disruption to GNSS*, London Economics. https://assets.publishing.service.gov.uk/government/uploads/system/uploads/attachment_data/file/619544/17.3254_Economic_impact_to_UK_of_a_disruption_to_GNSS_-_Full_Report.pdf (archived at https://perma.cc/7P33-KJ7A)

SAFETY4SEA (2018a) Stena Line to test AI technology onboard ship. https://safety4sea.com/stena-line-to-test-ai-technology-onboard-ship/ (archived at https://perma.cc/2TLM-272D)

SAFETY4SEA (2018b) New Big Data project to enhance safety at sea. https://safety4sea.com/new-big-data-project-to-enhance-safety-at-sea/ (archived at https://perma.cc/5AQV-C5HK)

The Maritime Executive (2017) GPS spoofing patterns discovered. www.maritime-executive.com/article/gps-spoofing-patterns-discovered (archived at https://perma.cc/9MW9-38FA)

United Nations Conference on Trade and Development (2018) *Review of Maritime Transport 2018*, UNCTAD. https://unctad.org/en/pages/PublicationWebflyer.aspx?publicationid=2245 (archived at https://perma.cc/4X4V-UMTL)

Wall, R (2018) Airbus looks windward, will put sails on ships moving plane parts. www.wsj.com/articles/airbus-looks-windward-will-put-sails-on-ships-moving-plane-parts-1536400806 (archived at https://perma.cc/5N5M-6LNY)

Wingrove, M (2015) Shipping slow to adopt condition based maintenance. www.marinemec.com/news/view,shipping-slow-to-adopt-condition-based-maintenance_41206.htm (archived at https://perma.cc/NUZ6-ZWMZ)

World Maritime News. (2018) Wärtsilä opens world's 1st international maritime cyber center of excellence. https://worldmaritimenews.com/archives/262652/wartsila-opens-worlds-1st-international-maritime-cyber-center-of-excellence/?uid=94645 (archived at https://perma.cc/3J6N-CDX9)

03

Ports and harbours

Leading the global charge against harmful emissions

As we have seen in Chapter 2, the world still relies heavily on seaborne traffic to carry out its daily business. Eighty per cent of world trade is carried by sea, and the essential conduit enabling that activity remains the global network of ports and harbours, many of which have been operating in some shape or form for millennia. The UK ports sector is one of the largest in Europe, in 2017 handling some 481.8 million tonnes of freight across all of its ports (Department for Transport, 2018a), and 21.5 million international sea passengers (Department for Transport, 2018b).

The UK port sector comprises 40 major commercial port cities and towns, which handle 70 per cent of the UK port tonnage. These include major all-purpose ports such as Southampton, London and Liverpool; ferry ports such as Dover; specialized container ports such as Felixstowe; and ports catering for specialized bulk traffic, such as coal or oil. These major ports are complemented by approximately 80 smaller commercial ports that may cater for local traffic or specialize sectors such as fishing or leisure boating.

Over 95 per cent of UK imports and exports by volume pass through the country's ports. It is estimated that in 2015 UK ports directly contributed approximately £22.6 billion in business turnover, £7.6 billion in gross value added and 101,000 jobs to the UK economy (Centre for Economics and Business Research, 2017). Internationally, it is estimated that 839 major ports handle about

99 per cent of global trade, with many thousands of smaller ports handling the remaining 1 per cent.

In addition to being important modal hubs in a country's transport system, many ports are also centres of local economic activity. Industries such as oil refineries and power stations as well as a range of businesses are in or near ports. Ports themselves are increasingly diversifying their activity into logistics and other value-added services.

Ports in a storm

When you understand quite how important they are to global trade, it is no surprise that the world's ports and harbours have to consider multiple risk factors when it comes to security issues – including potential terrorist attacks on critical infrastructure, people smuggling, and low-level or industrial-scale theft from ships or from containers on land. Aligned to many other initiatives emerging in the Smart Port City movement, a range of tech-enabled security options are being trialled to counter these threats. One of latest is a Dutch–Belgian project looking to harness the Internet of Things (or the 'Internet of Tarpaulins', as one report named it) (Vleugels, 2018). A range of smart sensors aims to notify security when they think, among other things, that potential intruders are suspected to be climbing a fence, or potential thieves or smugglers are attempting to cut through tarpaulins. Image analysis software is also in development.

The use of security cameras on drones has been in development for some time within the port and harbour environment. As far back as 2014, the Abu Dhabi Ports Company has been utilizing remote-controlled flying drones to patrol its operations in Khalifa Port, Zayed Port, the Free Port and the New Free Port (The National, 2014). From the start, the 'eye in the sky' cameras utilized high-definition video and 14 megapixel still photographs, with the ability to transfer data to operators shoreside via its own Wi-Fi network.

The technical dexterity of drones, also known as unmanned aerial vehicles (UAVs), has progressed significantly since then, and they are now being used in more complex monitoring and evaluation tasks within the port environment. In 2018, for example, the Port Hueneme Division of the US Navy's Naval Surface Warfare Center announced a new partnership with UAV provider Aerial Alchemy to use drone-delivered imagery to help assess the maintenance needs and readiness of the US fleet. The aerial drone takes the images it captures of the exterior surface of naval ships and develops 3D digital models that can enable the identification of damage, corrosion and alignment issues (Rees, 2018).

Ports and harbours also have multiple reasons to want to know what lies beneath the waves under their control. Sadly, in recent years, attacks on ships in harbour and in maritime straits in areas of poor governance – most notably those on the *USS Cole* and the *MV Limburg* – have revealed the growing scale of threat from seaborne terrorists, some of whom have selected subsurface attack operations as their preferred method of attack. For example, after Sri Lanka's civil war came to an end, crude sea submarines being fashioned by the Tamil Sea Tigers were discovered in the jungle (Venkataramanan, 2009).

Terrorist groups have been seen to demonstrate attack capabilities using retrofitted aerial drones, so any port authority must assume that the development of subsea aquatic drones is also in consideration to widen the terrorists' capabilities and opportunities for attacks on coastal cities. Perhaps the most immediate threat – which has a long history of success – is that of mines or improvised explosive devices that can be laid from a wide variety of commercial and recreational vessels. These devices might come from redundant or surplus military stocks available in a global black market or be underwater improvised explosive devices that can be laid on the seabed or attached to structures and vessels.

As well as the hazards from explosives, the underwater threat posed by smugglers and traffickers seeking lucrative access points to ports and harbours is growing. It is already, for example, relatively common practice for drugs, illegal materials and arms to be smuggled in externally mounted pods ('parasites') below the waterline of ships' hulls. The examination of hulls before arrival in port or while in harbour is a time-consuming and operationally restricting exercise, often involving excessive and commercially prohibitive delays and the use of divers in near-zero visibility and sometimes dangerous conditions.

Beyond direct attacks, and as you would expect as a matter of course further out to sea, regular sub-surface port and harbour inspections are also required to assess critical infrastructure and ensure that changes in the port's seafloor have not been so severe that remedial actions are required. Most port authorities also have a statutory requirement to find and 'mark' the sites of new wrecks or other environmental hazards so that appropriate messaging can be sent to local mariners to ensure safety of navigation.

Autonomous systems – being put to good use in so many areas further out to sea – also offer great promise in being able to monitor and protect critical infrastructure, undertake underwater structure inspection and facilitate seafloor survey and change detection analysis in ports, harbours and their surrounding waters

However, the debate around security and autonomous vessels within the port environment is currently more concerned with such vessels being an additional security concern, rather than adding a layer of comfort. As the shipping industry continues to get to grips with how the growing threat of cyber security could affect its operations in years to come, it is perhaps understandable that some scepticism may prevail when considering the introduction of fully autonomous systems or those that can be controlled remotely via satellite communications – all with underpinning systems that could, in theory, be hacked.

Still, forward thinking ports are engaging with the autonomous agenda. In April 2018, the Maritime and Port Authority of Singapore announced a new memorandum of understanding with Keppel Offshore & Marine and the Technology Centre for Offshore and Marine, Singapore that paved the way for the joint development of autonomous vessels to be used within and around the port. Potential activities that such vessels could carry out within the harbour environment were given as channelling, berthing, mooring and towing operations (Shi Wei, 2018). Note also the active trials in the Port of Antwerp of the Echodrone, which is explored in greater detail in Chapter 11 on subsea monitoring (Maritime Cyprus, 2018).

Safe berth

Automation takes on wider connotations within the port and harbour environment. As major logistics hubs, ports are not only concerned with maritime traffic in general, but also the specifics of what goes where and when. Automated or semi-automated services are beginning to show real promise here, and one area of great interest involves the potential for self-docking.

The general public were fascinated with the very idea of self-parking cars when videos of early demonstrations at trade shows as far back as 2006 started to be released (Hawkins, 2010). What is less well known is that similar concepts are at an advanced stage within the shipping logistics sector. A significant breakthrough came in late 2018 when the Norwegian roll on, roll off passenger ferry *Folgefonn* completed a successful automatic docking exercise (Jallal, 2018). Once the human operator had selected the destination berth, the 85-metre-long fully electric vessel left the dock, manoeuvred out of the harbour autonomously, and sailed all the way to the next port of call. Once there, it also manoeuvred through the harbour entrance, and docked alongside the terminal. All of this took place without human intervention, with navigation dictated by a series of pre-determined tracks and waypoints.

Not surprisingly, the Norwegian Maritime Authority were fully involved in the trial and announced that they were impressed with the smoothness of the demonstration. This was the first time that a vessel of this size had completed a fully automated dock-to-dock operation, free from human intervention. Mooring operations are a particularly risky stage of overall port operations and, sadly, often result in accidents to both ship and shore crew. As highlighted in Chapter 12 on the safety of life at sea, such accidents can be fatal (SAFETY4SEA, 2018). While much more testing is obviously still required, smoothing auto-docking operations has the potential to remove human error and thus increase the safety and overall efficiency of the docking and undocking operations for ships, which happens many thousands of times a day around the world.

The autonomous journey does not stop there, however. Once ships are safely docked and their cargo has been unloaded, it then needs to be transported to its final destination, and that journey can often continue by water. In January 2018, Dutch manufacturer Port-Liner announced plans for 52-metre barges for such onward operations (Boffey, 2018). From summer 2018, these fully electric – and potentially crewless – container barges began to operate from the ports of Antwerp, Amsterdam and Rotterdam.

The first batch of five barges – each capable of travelling for 15 hours at a time and carrying 24 by 20 foot containers – can operate without any crew. As with many autonomous innovations, however, the rest of the infrastructure must be in place, too, so in the first instance they were operated manually. Such potential is good news for cost efficiency and, of course, encouraging news for the planet, as these vessels are entirely emission-free. Designed for the inland waterways of Belgium and the Netherlands, this new breed of vessel plans to take a disruptive tilt at the vast number of diesel-powered trucks currently moving freight in these areas, with 23,000 trucks estimated to be taken off the roads as a result. It is hoped that larger vessels (over twice as long and capable of travelling for 35 hours) will follow later. The manufacturers claim that their use alone could lead to a reduction of about 18,000 tonnes per year of carbon dioxide.

Something in the air

The immediate environmental benefits to be delivered by these innovative electric barges are very much to be welcomed, but on their own they will not be able to tackle one of the major issues facing ports and harbours and, indeed, many other industries and citizens.

Poor air quality has been identified as a major problem globally, for the United Kingdom as a whole, and for ports in particular. Globally, nine million people died in 2015 as a result of air pollution (Griffin, 2017), and the problem is estimated to lead to 50,000 deaths a year in the UK, accounting for some 8.39 per cent of total fatalities (Silver, 2017). Pollution-related diseases account for an estimated 1.7 per cent of healthcare spending in high-income countries such as the UK. The problem is estimated to cost the UK economy alone some £20 billion a year. Air pollution is linked to major health issues – stroke, heart disease, dementia, asthma, lung cancer, chronic obstructive pulmonary disease – and has been shown to have life course development impacts.

A problem as broad, pronounced and entrenched as this usually inspires commercial interest. Increasing adoption of air quality monitoring stations in different industries is anticipated to fuel the growth of the global air quality *monitoring* market, which is expected to be valued at £7.46 billion by 2025 (from £3.3 billion in 2017), at a compound annual growth rate of around 8.31 per cent between 2018 and 2023 (Credenceresearch.com, 2018). Additionally, the global air pollution *control systems* market is expected to be valued at £77.15 billion by 2025, registering a 5.0 per cent compound annual growth rate over the period (Prnewswire.com, 2018). The UK market was valued at £2.17 billion in 2017.

Forty-seven towns and cities in the UK are at or have exceeded air pollution limits set by the World Health Organization. Most importantly, as this major issue relates to the Blue Economy, nearly a third of those are in or are adjacent to port areas (BBC News, 2018).

The major factors driving the air quality business growth outlined above include increasingly supportive government regulations for effective air pollution monitoring and control; ongoing initiatives towards the development of environment-friendly industries; increasing public–private funding for effective air pollution monitoring; rising levels of air pollution; and increasing public awareness related to the environmental and healthcare implications of air pollution. Each of these issues also applies to the ports and harbours of the world.

With statistics as well-understood and challenging as this, it is no surprise that the policy and regulatory framework is also trying to introduce pressure to bring about positive impact. For example, the European Union is imposing potential fines on the UK of £259 million for failure to meet air quality targets and the UK Government is now committing significant sums to support authorities in the implementation of both clean air strategies and implementing entire clean air zones, with charging being introduced for freight movements from 2020 to meet air quality targets.

All these factors are very relevant to the world's major ports and harbours, as they are the focal point of so much trade. What can be harder, however, is how to work out who the worst perpetrators are, and therefore how best to bring about new measures to make a positive difference. Within the complex operating environment of a major port, air quality risks are currently hard to attribute; baseline data are often inadequate and of poor quality, with the constant concern that impacts are associated with the wrong causal factors (eg port activities versus traffic congestion and their relative contribution to air pollution), interrelationships are poorly understood, and mitigation techniques are not quantifiable, effective or provable.

This can lead port operators to be targeted as significant contributors to air pollution levels based on sketchy evidence. Port operators and the cities they are attached to are also facing challenges of meeting more stringent global emissions controls and regulations (Sulphur 2020, clean air zones, etc), with concerns over implementation and monitoring. The IMO will restrict all shipping fuel oil to no more than 0.5 per cent of sulphur by 2020. As we also discuss in Chapter 5, cruise and ferry companies are also facing more and more discerning customers where the impact of their journeys and the environmental impacts matter greatly to them, and there is increasing pressure from the governments responsible for protected or environmentally vulnerable areas. Air quality and managing emissions is undoubtedly a problem that is not going to go away any time soon. So, what is being done within the port environment?

On the global stage, the US Port of Los Angeles is offering bold leadership on the issue, despite the complexities outlined above. In August 2018, the port showed that reducing emissions does not have to have a negative impact on revenues or trade output (a fear sometimes cited by port-based companies and port authorities themselves) when it announced that it had managed both to establish new record lows for emissions reductions within its associated operations at the same time as driving container volume to an all-time high of 9.34 million TEU (World Maritime News, 2018a). The port's official Inventory of Air Emissions for 2017 recorded that emissions of nitrogen oxides were down some 60 per cent compared to emissions levels recorded in 2005, their lowest level to date. The broad findings within the report showed that the port had maintained or exceeded the clean air progress it had made over the previous decade, meaning that it met all of the goals it had set within its 2023 Clean Air Action Plan – some six years ahead of time.

The Port of Rotterdam set similarly ambitious targets to help meet its commitment to reducing carbon dioxide emissions from shipping and industry – aiming to slash emissions by 49 per cent by 2030 and 90 per cent by 2050 (gCaptain, 2018a). Achieving these figures would smash the targets set by the IMO in April 2018, to reduce emissions by at least 50 per cent by 2050.

Such gains cannot be made without significant financial investment. As one example of the levels of support needed, in early 2019 the Port of Long Beach announced the release of $147 million to support six new projects that would develop zero-emissions equipment and advanced operational energy systems (Nahigyan, 2019). Again, these projects – which drew on both public and private funds – were aligned to the same Clean Air Action Plan that has been guiding the Port of Los Angeles towards zero-emissions cargo-handling equipment by 2030 and zero-emissions trucks by 2035.

Funds drawn from such sources as the California Energy Commission and the California Air Resources Board were invested in projects that included a large-scale demonstration of zero-emissions cargo handling equipment involving electric gantry cranes and tractors, amongst other equipment, and an ambitious proposal to develop an electricity microgrid.

In all ports, employing a wide range of innovative technologies will be essential to meet such targets. It was reported recently that ship owners are scrambling to install scrubbers, which filter sulphur from dirtier fuel oil, to help meet 2020 emissions targets (Bousso and Ghaddar, 2018), and Rotterdam is helping to nudge ship owners further in the right direction by providing financial incentives for low- or zero-carbon vessels. Their Environmental Ship Index began measuring the emissions of individual ships last year, and they have also launched a digital platform where shipping companies and service providers can exchange information about their port visits in order to increase efficiency in port – an initiative that could on its own cut emissions by up to 20 per cent.

Taking down the particulates

One award-winning innovation trying to solve this conundrum is a UK company called Green Sea Guard. Winner of the 2018–19 Rushlight Environmental Analysis and Metrology Award, the Green Sea Guard system is a device that measures and remotely tracks ship exhaust emissions of gases and particulates. Today, 75 countries representing approximately 90 per cent of world shipping have signed up to the MARPOL standards (the International Convention for the Prevention of Pollution from Ships). Specific emissions control areas have also been established where all traffic must comply; sulphur oxide levels are already controlled, and nitrogen oxide control areas are under discussion. There is also some demand for detailed carbon dioxide emissions data.

While all of this is of course to be warmly welcomed, proving compliance is not as easy. While there are more than 115,000 ships to monitor, port authorities and coastguards have no specific and standardized tool set that allows them to track emissions without physically boarding the ships. As a rough rule of thumb, it is estimated that most coastguards and port authorities understandably do not have sufficient resources to inspect more than one ship per thousand of the shipping traffic passing through their areas. Well-meaning regulations therefore may not be as great a deterrent as intended to ships that do not comply with international emissions standards. With such limited enforcement resources, offenders know that the probability of being caught are small, even though shoreside authorities do have the power to detain non-compliant ships.

The warning signs are certainly there for ship owners attempting to flout the rules. In 2018, the Mediterranean Shipping Company SA paid $630,625 in penalties to the California Air Resources Board for violating its Ocean-Going Vessel At-Berth regulation (gCaptain, 2018b). These violations were discovered during a routine audit of the company's 2014 visits to the Port of Oakland and the twin ports of Los Angeles and Long Beach. Over 2,500 violations were identified – for failing to reduce auxiliary engine power generation by at least 50 per cent and for exceeding limits for auxiliary engine run time. The California Air Pollution Control Fund was the recipient of the fine and earmarked it to undertake additional research into air pollution.

While all of this paints a troubling picture, any emerging business has to find a positive angle that brings quantifiable value to all potential customers. With budgets for regulators and port authorities limited, Green Sea Guard aims to do so by promoting the impact of transparent compliance on emissions levels. Compliant ship owners will have already invested in a range of processes to ensure the fall in line with regulations regarding low emission engines and fuel. However, if port authorities cannot see that, they do not know *not* to target their vessels for necessary compliance checks.

Many regulations are enforced by manual inspection of fuel tanks and by inspecting paper fuel receipts. At best, these methods are time-consuming, which can impact the profitability of ships that are pulled in to undergo such procedures. Green Sea Guard targets this pressure point as the value proposition to the ships they are providing a service to. Manual compliance tank inspections can take up to four days to complete, during which time the ship will be incurring additional operating cost and significant delay to schedules.

Some port authorities are also exploring charging differing port fees for highly and less polluting vessels. The Port of Gothenburg, for example, operates an environmental discount on port tariffs, intended to reward those ships that comply with approved standards (Portofgothenburg.com, 2019). Currently, though, to achieve this discount, ship owners need to prepare and send notifications of registration and compliance with approved standards to the Gothenburg Port Authority. A fully adopted Green Sea Guard system could remove the need for such paperwork, if emissions data is available to all authorities via transparent digital access, thus enabling them to automate and significantly increase the scale of their inspection capability.

Green Sea Guard's units are compact pieces of equipment that plumb directly into the engine exhaust in the engine room (Burgess, 2018). It is then able to monitor the exhaust flow rate and temperature. It will sample the exhaust at user-set intervals that can range from seconds to minutes. The system stores the results locally, undertakes some basic analysis onboard ship and transmits the records to the secure server on land. Ship owners, port authorities and coastguards can access the data securely (from a screen based in their offices on land or at sea) using the Green Sea Guard portal and can track each vessel that is of interest to them. This toolset allows coastguards and port authorities to track and monitor all ships. This ability is a significant improvement on random spot checks at bridges using sniffer technology, or by helicopter.

Setting the scene for innovation

The Port of Rotterdam in the Netherlands, the largest in Europe, has publicly declared that it wants to be the 'smartest' port in the world (World Maritime News, 2018b). The range of innovation programmes and high-level partnerships it runs across its 42-kilometre site is undoubtedly impressive. For example, the Internet of Things platform that it is developing in partnership with IBM, Cisco, Esri and Axians utilizes a broad network of sensors to provide accurate and up-to-date water (hydro) and weather (meteo) data to help the port authority to plan and manage shipping operations more effectively (Porttechnology.org, 2019). They hope that this system will lead to decreased waiting times, optimized berthing and faster loading times.

The port is also exploring the potential of blockchain technology. This digital distributed ledger system is being hyped across many industries, but there is sound logic for its use within the port environment to help make logistics processes more efficient. For example, the extensive paperwork and consignment notes that accompany the many thousands of containers moved by the global shipping industry daily are ripe for new, digital solutions. Blockchain is being explored in Rotterdam and in other leading ports globally, and looks especially promising as it has the potential to flatten out multiple registration and control processes that can involve up to 25 separate entities in relation to a single transport transaction (eg the port authority itself, customs agents, stevedores, freight forwarders, road carriers, shippers, etc).

To add further energy to the search for solutions to the air quality and emissions issue, the Port of Rotterdam also announced in January 2019 that it was inviting applications to a funding pot of €5 million designed to promote climate-friendly shipping (SAFETY4SEA, 2019). This scheme – which will run until 2022 – will support ship owners, charterers, fuel suppliers and producers to test low-carbon or zero-carbon fuels – particularly those that can help cut the sector's carbon dioxide emissions by over 50 per cent.

Rotterdam was the site of the world's first specialist Port and Maritime Accelerator, known as PortXL. Launched in 2015, in selecting its latest round of successful applicants the operation states that it has scouted 1,000 start-ups, and accelerated 36, which has led to the signing of over 80 pilot contracts (PortXL, 2019). The PortXL concept is now being established in Singapore, Antwerp and promises to launch in the USA next.

As well as focusing on what may be seen as the excitement of being involved in the development of innovations such as blockchain and autonomous and remotely operated vessels, the Port of Rotterdam is also focusing its energies on some of the 'back office' functions and tasks that will make further progress in all these areas more efficient. For example, it was very encouraging to see Rotterdam announce in November 2018 that an important step forward in the standardization of data across ports had been agreed (Porttechnology.org, 2019).

It is an oft-heard complaint from digital innovators that one of the main barriers to achieving global scale is having to develop systems that attempt to achieve the same tasks in several international territories but must tackle completely different approaches to and formats of data in each region. Addressing this issue is also one of the main areas of interest of organizations such as the UK's Open Data Institute, which understands that promoting common data standards will make it much easier for new data-driven services to be developed

(Dodds, 2018). Back in Rotterdam, the November 2018 announcement focused on the establishment of common nautical standards for port information that will, for example, enable terminal operators to share a vessel's berthing details with multiple parties in order to improve port call efficiency.

The digitalization agenda thrives on data, so such standardization initiatives are hugely important (even if they seem a little dry at first) as they provide the platform for the next wave of innovation.

Conclusion

Ports and harbours make very interesting test beds for technologies that may eventually have much broader aspirations. They present opportunities to experiment with technologies that engage with the global shipping industry; that look to protect and improve infrastructure; that harness Big Data and satellite capabilities – all within relatively safe confines with decent connectivity, protection against the harsher weather elements and in close proximity to shore-based recovery and repair facilities.

As well as promoting innovation and experimentation, ports and harbours can act as useful focal points to consider and tackle legal questions. Quite often, when really innovative new technology comes to market, the specific regulations required to allow them to operate safely don't exist in law, so many of the leading ports have lawyers working with government to explore issues and make things possible.

Quite often, the port authorities themselves do not need to be developers of technology. Their aim is to create the right environment for innovation to flourish, and when systems subsequently become market ready, they may be ready to use them, for the good of their own customers, their broader port stakeholder community and so that they can meet the targets to which they themselves are held.

The scale of ambition is to be lauded. For example, the Internet of Things platform being developed by the Port of Rotterdam involves the placing of sensors everywhere from quay walls, to dolphins, waterways and roads. Therefore, as well as facilitating the smooth passage of global trade, Rotterdam's infrastructure is now constantly generating measurement data that can be collected and analysed and quickly communicated to other autonomous systems to help drive ever higher levels of efficiency. With up to 1.2 million data points being processed every day, when aligned to the greater push towards standardization of nautical data as it relates to ports, this may be advanced, but it is clearly just the beginning of the smart port city.

References

BBC News (2018) UK's most polluted towns and cities revealed. www.bbc.co.uk/news/health-43964341 (archived at https://perma.cc/9GKD-KNXW)

Boffey, D (2018) World's first electric container barges to sail from European ports this summer. www.theguardian.com/environment/2018/jan/24/worlds-first-electric-container-barges-to-sail-from-european-ports-this-summer (archived at https://perma.cc/9WEX-W7CM)

Bousso, R and Ghaddar, A (2018) Ship owners are scrambling to install scrubbers. www.marinelink.com/news/ship-owners-scrambling-install-scrubbers-442681 (archived at https://perma.cc/YN84-EFKN)

Burgess, A (2018) Green Sea Guard: Supporting energy transition in the shipping industry. www.governmenteuropa.eu/green-sea-guard/90480/ (archived at https://perma.cc/22J3-X4YG)

Centre for Economics and Business Research (2017) *The economic contribution of the UK ports industry: A report for Maritime UK*. www.maritimeuk.org/documents/187/Cebr_Ports_report_finalised.pdf (archived at https://perma.cc/9M77-63P6)

Credenceresearch.com (2018) Air quality monitoring equipment market to reach US$ 9.5 bn by 2025. www.credenceresearch.com/press/global-air-quality-monitoring-equipment-market (archived at https://perma.cc/GC6Z-GJ3E)

Department for Transport (2018a) *UK Port Freight Statistics: 2017*. https://assets.publishing.service.gov.uk/government/uploads/system/uploads/attachment_data/file/762200/port-freight-statistics-2017.pdf (archived at https://perma.cc/GE26-5D3V)

Department for Transport (2018b) *Sea Passenger Statistics: All routes 2017 (final)*. https://assets.publishing.service.gov.uk/government/uploads/system/uploads/attachment_data/file/754201/sea-passenger-statistics-all-routes-final-2017.pdf (archived at https://perma.cc/P69S-95N4)

Dodds, L (2018) Documenting the development of open standards for data. https://theodi.org/article/documenting-the-development-of-open-standards-for-data/ (archived at https://perma.cc/2MQW-CG85)

gCaptain (2018a) Zero carbon at sea? Rotterdam Port eyes a greener future. https://gcaptain.com/zero-carbon-at-sea-rotterdam-port-eyes-a-greener-future/ (archived at https://perma.cc/PCJ3-ULQQ)

gCaptain (2018b) MSC pays $630,000 in penalties for California air quality violations. https://gcaptain.com/msc-pays-630000-in-penalties-for-california-air-quality-violations/ (archived at https://perma.cc/K2QP-68WN)

Griffin, A (2017) The truth of air pollution was just revealed. And it is horrifying. www.independent.co.uk/environment/pollution-air-clean-water-vehicles-diesel-car-tax-lancet-report-deaths-fatal-disease-a8009751.html (archived at https://perma.cc/Y848-8P97)

Hawkins, L (2010) The skinny on self-parking. www.wsj.com/articles/SB10001424052748703734504575125883649914708 (archived at https://perma.cc/E3ZK-WW6U)

Jallal, C (2018) Wärtsilä achieves notable advances in automated shipping with latest successful tests. www.marinemec.com/news/view,selfdocking-vessels-sail-one-step-closer-to-berthing_56055.htm (archived at https://perma.cc/5KS2-3TY8)

Maritime Cyprus (2018) A first for the port of Antwerp: Innovative autonomous sounding boat with unique technology. https://maritimecyprus.com/2018/08/17/a-first-for-the-port-of-antwerp-innovative-autonomous-sounding-boat-with-unique-technology/ (archived at https://perma.cc/QLY6-UL47)

Nahigyan, P (2019) Grant funds help port prepare for transition to zero-emission tech. www.lbbusinessjournal.com/grant-funds-help-port-prepare-for-transition-to-zero-emission-tech/ (archived at https://perma.cc/26R8-DUDK)

Portofgothenburg.com (2019) Environmental discount on the port tariff. www.portofgothenburg.com/about-the-port/greener-transport/environmental-discount-on-the-port-tariff/ (archived at https://perma.cc/JK3Y-SQVT)

Porttechnology.org (2019) Port of Rotterdam unveils IoT breakthrough. www.porttechnology.org/news/port_of_rotterdam_unveils_iot_breakthrough (archived at https://perma.cc/DJG3-TRG3)

PortXL (2019) Vision. https://portxl.org/vision/ (archived at https://perma.cc/M9SZ-RDHH)

Prnewswire.com (2018) Global air pollution control systems market, 2018–2025: Market size is expected to be valued at USD 98.17 billion by 2025, registering a 5.0% CAGR. www.prnewswire.com/news-releases/global-air-pollution-control-systems-market-2018-2025—market-size-is-expected-to-be-valued-at-usd-98-17-billion-by-2025–registering-a-5-0-cagr-300718550.html (archived at https://perma.cc/74LP-2FTB)

Rees, M (2018) US Navy investigating drone-based maintenance services. www.unmannedsystemstechnology.com/2018/09/aerial-alchemy-provides-u-s-navy-with-drone-based-maintenance-services/ (archived at https://perma.cc/Y6B2-NXHN)

SAFETY4SEA (2018) Seafarer killed during container lashing in Port of Dublin. https://safety4sea.com/seafarer-killed-during-container-lashing-in-port-of-dublin/ (archived at https://perma.cc/5MV4-BGSU)

SAFETY4SEA (2019) Port of Rotterdam launches scheme to promote climate-friendly shipping. https://safety4sea.com/port-of-rotterdam-launches-scheme-to-promote-climate-friendly-shipping/?utm_source=noonreport&utm_medium=email&utm_campaign=other (archived at https://perma.cc/8W5U-Z3TK)

Shi Wei, N (2018) Singapore Maritime Week: MPA signs Tuas Terminal Phase 2 contract, autonomous vessels MOU and more. www.straitstimes.com/business/economy/mpa-signs-tuas-terminal-phase-2-contract-autonomous-vessels-mou-and-more (archived at https://perma.cc/WK42-X8NE)

Silver, K (2017) Pollution linked to one in six deaths. www.bbc.co.uk/news/health-41678533 (archived at https://perma.cc/F9GN-ZAPG)

The National (2014) Eye in the sky: Abu Dhabi's ports now protected by drones. www.thenational.ae/business/eye-in-the-sky-abu-dhabi-s-ports-now-protected-by-drones-1.595388 (archived at https://perma.cc/EG9S-TXRT)

Venkataramanan, K (2009) Forces find 'submarines' in abandoned LTTE base. https://timesofindia.indiatimes.com/world/south-asia/Forces-find-submarines-in-abandoned-LTTE-base/articleshow/4048344.cms (archived at https://perma.cc/27YK-N7CR)

Vleugels, A (2018) How to catch criminals through IoT and predictive software. https://thenextweb.com/the-next-police/2018/11/01/police-iot-ports-crime/ (archived at https://perma.cc/2KAW-4UHT)

World Maritime News (2018a) Los Angeles Port sets new record for cutting NOx emissions. https://worldmaritimenews.com/archives/259415/los-angeles-port-sets-new-record-for-cutting-nox-emissions/?uid=94645 (archived at https://perma.cc/LX2Y-S8ZJ)

World Maritime News (2018b) Port of Rotterdam wants to be the smartest. https://worldmaritimenews.com/archives/254068/port-of-rotterdam-wants-to-be-the-smartest/ (archived at https://perma.cc/ZBG3-Y86A)

04

Offshore renewables

Channelling the power of the oceans

As landmarks go, it was undoubtedly a significant one, and certainly a long time in the making. On 6 November 2018, the United Kingdom was able to tell the world that the electricity output from its renewable energy sources in the previous quarter had overtaken that produced by fossil fuels (Vaughan, 2018). So, for the very first time, the combined 42 gigawatt capacity of wind, solar, biomass, hydro and other forms of renewable energy had outperformed the 40.6 gigawatts generated by fossil fuel generators such as gas and coal. The statistics, provided by Imperial College London, were in stark contrast to the status quo at the start of the decade, when generating capacity from coal, oil and gas was seven times greater than that of renewables.

While this only represents one small step in a long journey, any progress in this field is to be heartily welcomed, for multiple reasons. In the United States, while the Department of Energy reports that the country generated approximately 37 per cent of its electricity from zero-carbon sources in 2018 (20 per cent nuclear; 10 per cent wind and solar; 7 per cent hydropower), that still leaves 63 per cent generated from non-renewable sources, which, as we shall see, looks environmentally unsustainable (Eia.gov, 2019).

This book is packed with exciting new uses of digital technology. Huge new data stores teem with new Earth observation information being pored over by scientists, technologists and students. High-powered computing, machine learning and artificial intelligence systems enable calculations to unprecedented levels. The feature-packed and sensor-laden possibilities of the near-ubiquitous smartphone and other mobile devices are revolutionizing the at-sea collection, management and analysis of data. Wrist bands use GPS technology and embedded networks of sensors to let holidaymakers know where they are and provide new seamless ways of booking trips, ordering drinks and opening doors. While for the Blue Economy this is encouraging and exciting in equal measure, the environmental flip side of all this amazing technology is that they all significantly increase the demand for one thing – power. And that is a trend playing out across all industries.

Global energy consumption has more than tripled in the past 50 years, in the main because of rapid economic and population growth in the Asia Pacific region, trends that look set to continue. According to the International Energy Agency, the continued rise in incomes and a further population increase of 1.7 billion people will make global energy demand soar by an additional 25 per cent by 2040 (Iea.org, 2018). This aligns to the macro shifts being witnessed between East and West. For example, in 2000, North America and Europe jointly accounted for more than 40 per cent of global energy demand, with developing economies in Asia responsible for roughly half that. This situation is set to reverse totally by 2040.

Southeast Asia consists of 11 countries with a total combined population of roughly 630 million, a number that is expected to increase by a further 25 per cent by 2050. Energy consumption in the region nearly doubled in just over two decades. Around 65 million people in the region already lack adequate or reliable electricity access, and energy security concerns are rising further both because demand is expected to grow by an average of 4.7 per cent per year to 2035 and because plans need to be made to cope with the inevitable depletion of indigenous fossil fuels. Renewable energy delivered 17 per cent of the Southeast Asia's total electricity generation in 2015. Seventy-five per cent of that amount was provided by hydroelectric power, but non-hydropower renewables installed capacity also more than doubled in a decade, from 6 to 15 gigawatts. Much of this progress came about due to the £27 billion that was invested in renewables in Southeast Asia between 2006 and 2016, according to the International Renewable Energy Agency (IRENA), which also reports that, at the policy level, the 10 nations that make up the Association of Southeast Asian Nations (ASEAN) have jointly agreed on a 23 per cent target for sustainable, modern renewables by 2025. So, the region cannot be accused of not taking the issue seriously.

While more can always be done, as in every part of the world, it is also acknowledged that deploying renewable energy solutions helps the region to meet UN Sustainable Development Goals (SDGs). While renewable options are most obviously aligned to SDG7 (ensuring access to affordable, reliable, sustainable and modern energy for all), they also have cross-cutting relevance to goals focusing on climate change, health improvement, poverty reduction and access to clean water and nutrition.

Going down a level, the example of Indonesia illustrates why renewable energy will have such a critical role to play in the region. Indonesia, a member of the G-20 group of developed nations, is the largest economy in Southeast Asia. With a population of around 260 million, the country is the fourth most populous nation on Earth, served by the world's tenth largest economy. Electricity demand is increasing year on year.

While network capacity is expanding rapidly, providing reliable connection to rural and remote areas is costly and complicated. Diesel generators are still used in most villages and the country experiences regular shortages in supply, with power blackouts a regular part of life. Its need for more power also sits alongside another pressing short-, medium- and longer-term Blue Economy demand. The call for efficient renewable energy production also aligns to one of the issues that affect all chapters of this book in various ways – that of climate change.

A climate of fear

Over 40 million (approximately 15 per cent) of Indonesia's citizens live in low-elevation coastal zones. These regions are the most vulnerable to natural disasters and are areas where risk factors are further exacerbated by the effects of climate change, without even considering the potentially devastating effects of sea rise on these communities and the environments that currently sustain them. With more than 25.9 million Indonesians (approximately 10 per cent of the population) already living below the poverty line, the country needs to act urgently to minimize any further threats. In 2012, though, Indonesia was the sixth largest emitter of carbon dioxide in the world.

No wonder, then, that the Indonesian Government has committed to reduce emissions by 29 per cent by 2030, with a 'stretch pledge' of 41 per cent emission decrease said to be possible with international help. As energy use currently accounts for 25 per cent of all Indonesia's harmful emissions (and by 2030 it is predicted to be the biggest source), Indonesia has set itself a target for new and renewable energies to increase to 23 per cent (18 per cent renewable and 5 per cent other new energies) by 2025. To understand the scale of this ambition, and how challenging it will be to bring to fruition, achieving this target would entail an eleven-fold increase in renewable energy output from 2014 levels. Indonesia has a 'secret weapon' here because – nestling as it does within the Pacific 'Ring of Fire', a meeting point of four tectonic plates – it is estimated to possess 40 per cent of the world's potential geothermal resources. This already makes it the third largest geothermal electricity producer after the United States and its neighbour the Philippines, and it has announced plans to invest further in this area to become the world's undisputed producer of geothermal energy, eventually accounting for 5 per cent of Indonesia's total energy needs. This field is attracting such great interest because even with, for example, the World Bank investing heavily in the development of geothermal energy in the country in recent years, it is estimated that over 90 per cent of Indonesia's geothermal resource remains untapped.

However, other Indonesian energy sources appear to remain underdeveloped, too, not least when you consider the fact that the country also has approximately 34,000 miles of coastline. Leaving offshore wind aside for now, marine renewable energy possibilities broadly fall into two main categories: tidal energy (harnessing the kinetic energy found in the movement of large bodies of water) and wave energy (doing the same at the water's surface). Both can be used to drive turbines in order to produce electricity, and are renewable sources as they will be available for as long as the tides continue to ebb and flow, unlike the finite (and more harmful) options provided by fossil fuels. It should be noted that differences in temperature and salinity that occur in the ocean also present renewable possibilities, but this field is still very nascent.

Academics have recently been assisting in Indonesia's endeavours by analysing which part of the country's vast expanse of marine estate could be most appropriate for wave and tidal energy projects. As the country is made up of approximately 18,000 islands, the masses of narrow channels and straits in between archipelagos have the effect of amplifying currents, thus providing suitable locations for wind and tidal energy installations.

A research paper featured in IRENA's renewable energy roadmap for Indonesia in March 2017 (IRENA, 2017) that analysed estimates from 10 separate locations, for example, noted that the Alas Strait and the Riau Islands held nearly 70 per cent of the total tidal potential in the measured areas. Similarly, a paper in the *International Journal of Mechanical Engineering and Technology* in October 2017 (Sugianto et al, 2017) aggregated findings from 11 separate studies into the potential of wave power technology in various islands of the Indonesian archipelago. Even though both reports highlighted significant challenges related to the development of wave and tidal power in Indonesia (including querying the value of marine power as peak tidal speeds only last for a couple of hours a day, and pointing out the difficulties of developing supporting infrastructure), such data is undoubtedly nectar to renewable energy entrepreneurs, who – as we see in many cases – have developed technologies and need to understand in greater detail where best they may be deployed.

For example, the areas of greatest potential for wave power identified in the academic research cited above included Sumatera Island, South of Java and in the waters around Bali – which is where one such innovative renewable energy technology is attempting to enter the Indonesian market. In December 2017, Indonesian infrastructure construction company Gapura Energi Utama announced that it was establishing the world's largest wave energy park, to be located next to Nusa Penida Island, off the south-east coast of Bali. The provider of the technology is Finnish company Wello Oy, which will deploy its Penguin wave energy converter to provide 10 megawatts of power.

The Wello Penguin generates carbon-free power by harnessing rotational energy from waves (put simply: when it gyrates, wave power is converted) in waters up to 50 metres deep. The device needs to be anchored and transfers its electricity output to the grid via underground cables. Further good news for the local marine life comes with the fact that the system operates silently, and is built offsite and towed to its eventual location, so noise and construction pollution are minimized.

It has taken the Finnish co-founders Antii and Heikki Paakkinen over ten years to bring their new wave-converting device to market, with £6.4 million of investment being secured over the course of five funding rounds between 2009 and 2018. The Penguin was first fully tested in sea trials in 2012, safely handing rough sea states with waves up to 12 metres high. Such rigorous testing is crucial, even if it may lead to some setbacks in the early stages of development. Progress appears to have been smooth until, unfortunately, the company encountered a setback in March 2019 when one of its 1 megawatt Penguin devices sank off the coast of Orkney in Scotland. Two years after its initial deployment at the grid-connected test site of the European Marine Energy Centre, the unit was seen to be taking on water during an inspection, and then disappeared completely from view later in the week. While frustrating, such setbacks are not entirely unknown, and are a constant threat, especially when attempting to bring entirely new technology to bear in sea states that by necessity must be challenging – there is, after all, not much point in having a wave converter on pancake-flat seas!

On its website, Wello pitches itself against wind and solar energy by saying that 'the wind does not always blow, and the sun does not always shine'. Aside from those indisputable facts of life, and the ecological and commercial drivers behind their offer, the Wello Penguin and similar wave energy-converting technologies also have another key selling point in areas of natural beauty such as the island of Nusa Lembongan. As they are wave-level devices, they are far less visually intrusive than the more obvious offshore wind turbines that are now becoming a familiar sight in many countries' coastal areas, or even solar panels on land. Such installations can often attract strong opposition from local communities, an issue made even more pressing if they are in line of sight of known tourist attractions.

While for all the reasons stated above Bali needs to embrace renewable energy sources, it will not wish to do so in ways that threaten one of its main sources of revenue – tourism. In 2017, Bali on its own accounted for over five million of Indonesia's total of 14 million tourist visitors, so it is perfectly understandable that the governors of Nusa Lembongan island would not want to lessen the impact of its pristine white sand beaches or cliff viewpoints with tall man-made edifices, no matter how important they are to the island's sustainability and the broader reduction of climate change.

Not content with establishing the world's largest wave energy park, in 2018 Indonesia also announced that it was to build the largest tidal energy project yet attempted. A little under 400 miles east of Nusa Lembongan, plans are emerging in East Flores to construct a bridge in the Larantuka Straits. In the first instance, the 810-metre Pancasila-Palmerah Bridge will link the island of Adonara with its larger neighbour Flores. As well as being an extremely useful transport development, though, the bridge will also function as a dam, and five turbines will be installed to convert the rise and fall of the tides into enough electricity in its first phase to support 100,000 people. This first phase is expected to cost approximately £150 million, which will be met by private investment. Dutch consortium Tidal Bridge BV (a joint venture between Strukton International and Dutch Expansion Capital) received joint backing from, amongst others, Indonesia's Energy and Mineral Resources Ministry, the Public Works and Public Housing Ministry and the Ministry for Maritime Affairs, and it expects the bridge to start delivering electricity by 2020.

Critical factors for the siting of such projects are the height of the tidal range (the difference between the low and high points) and/or the tidal velocity (how fast the water flows). The tidal range in the Larantuka Strait is 3 metres, and its peak velocity has been measured at 3 to 4 metres a second, statistics that, when combined, make the project viable. One of the more attractive characteristics of this renewable energy process is that the regularity of the tides makes tidal power output more predictable than wave power, which can vary greatly due to weather conditions.

Winds of change

Even though it may be more variable, offshore wind is beginning to show great promise. The Global Wind Energy Council reports that the United Kingdom is the world's largest offshore wind market and accounts for just over 36 per cent of installed capacity, followed by Germany in the second spot with 28.5 per cent. China comes third in the global offshore wind rankings with just under 15 per cent. Denmark now accounts for 6.8 per cent, the Netherlands 5.9 per cent, Belgium 4.7 per cent and Sweden 1.1 per cent. Other markets include Vietnam, Finland, Japan, South Korea, the United States, Ireland, Taiwan, Spain, Norway and France.

A historical record of 4,331 megawatts of new offshore wind power was installed across nine markets globally in 2017. This represented an increase of 95 per cent on the 2016 market. At the end of 2017, nearly 84 per cent (15,780 megawatts) of all offshore installations were located in the waters off the coast of eleven European countries. The remaining 16 per cent is located largely in China, followed by Vietnam, Japan, South Korea, the United States and Taiwan.

The sector is enjoying especially unprecedented growth in the UK. The *Digest of UK Energy Statistics*, published by the Department for Business, Energy and Industrial Strategy, showed that in July 2018 offshore wind systems reached 17.2 per cent of the UK's total renewable capacity by type. This put it third in the ranking table behind onshore wind (with 31.7 per cent of total capacity) and solar photovoltaics (with 31.5 per cent), and ahead of bioenergy (with 14.9 per cent) and hydro (with 4.6 per cent). More impressive, though, was the pace of progress, with the report revealing that offshore wind generation had increased 27 per cent between 2016 and 2017. While some of this growth was attributed to the higher wind speeds experienced in 2017 compared to the previous year, this continued a steep rise that had seen the UK offshore wind sector increase its output by some 175 per cent since 2012.

The UK's *Offshore Wind Industry Prospectus* (Cdn.ymaws.com, 2018) estimates that by 2030 the global offshore wind sector could be worth as much as £30 billion to UK firms each year. This figure does not just include the offshore wind farms themselves, but also related supply chain services such as autonomous systems, sensor suites and data analytics, all of which are provided with new sales opportunities within this burgeoning market. Within such related services, the UK's Offshore Wind Industry Council suggests that – with the right support – UK companies could consolidate on their current position by providing 60 per cent of offshore wind farm turbines by 2030, up from the current figure of 48 per cent. While the UK is showing real leadership in this sector, it is not alone. Market research company Global Industry Analysis predicts that offshore wind capacity will grow by over 80 gigawatts by 2024. This means the sector will achieve a compound annual growth rate of more than 25 per cent over that period, so governments, investors and existing producers all appear to have great confidence in offshore wind's potential.

While initially lagging behind Europe, the offshore flame certainly looks to have been lit in the United States. Onshore and offshore wind generation is set to overtake hydroelectric power as its top source of renewable energy in 2019. Like elsewhere, this is just the start of the journey as the US's offshore wind potential is estimated to be greater than 22,000 gigawatts (Awea.org, nd), which is double the country's current electricity consumption. It is perhaps the pace of growth in the offshore wind sector that is most impressive, as it could still be seen to be an industry very much in its infancy, yet it is making huge strides. The Block Island Wind Farm off the coast of Rhode Island, for example, was the US's first ever offshore wind farm. It only started to operate fully in 2016, yet within one year it was already being hailed as a great success by analysts, politicians and local citizens alike. The five large 6 megawatt wind turbines that sit three miles off the island's coast are bringing electricity to the island that is better quality, more secure and cheaper. This means that the island's diesel generators, which were once a critical part of the energy mix, have been relegated to a back-up resource, only to be used in case of emergency. Such early positive results are driving a new appetite for offshore wind. In 2018 alone, the US offshore wind industry progressed from the single 30 megawatts Block Island offshore wind farm to a pipeline of at least 5 gigawatts.

In 2018, IRENA released global projections that growth in total installed offshore wind capacity will rise from 19.2 gigawatts in 2017 to 520 gigawatts in 2050 (IRENA, 2018), which if achieved equates to growth of a staggering 2,600 per cent over the period. This will be made possible by financial investments in offshore wind that will reach $350 billion by 2030 and $1.47 trillion by 2050. When the money starts to flow at this sort of rate (though predictions can obviously be erroneous), real progress at scale begins to look within grasp.

One of the more eye-catching reports in recent years related to offshore wind was produced by the Carnegie Institution for Science in Stanford, California. The headline-grabbing lead finding was that there is so much potential convertible wind energy gusting over the world's oceans that it could – in theory – generate what the authors called 'civilization scale power'. This meant that if it could be harnessed, wind power on its own could solve the world's energy crisis, fully replacing the use of harmful fossil fuels.

Wind speeds on the ocean have been measured at rates as much as 70 per cent higher than on land. Storms over the mid-latitude oceans regularly transfer wind energy down to the surface from high altitudes, making a much higher upper limit on how much energy wind turbines can capture than on land. The study also notes that wind energy gathered on land has a finite upper limit due to how structures on the land, both natural and manmade, can significantly interrupt and slow down wind speeds. Land-based turbines themselves slow the air, reducing the amount of energy subsequent rows of turbines can generate. The ocean, on the other hand, has a much higher limit.

The report's authors were quick to point out the limitations of their own headline, saying that they didn't think that the required development on such a scale to take full advantage of available capacity was likely. Investment cost aside, as above, not many coastal residents would vote to allow the installation of wind turbines as far as the eye can see. But the point has been reinforced that there is a huge amount of untapped energy in the oceans.

The main issue in accessing it at scale as envisaged here is that traditional offshore wind farms have needed to be close enough to shore that they can be securely anchored to the seabed. The further out to sea they are placed, the more complicated, expensive and dangerous that becomes, so fixed offshore wind installations have their limitations. However, whole new tiers of opportunity begin to open with the latest round of promising innovation in this sector – floating wind farms.

Floating a new idea

In October 2017, the Scottish grid started to take in energy from a wind farm 25 kilometres off the Peterhead in Aberdeenshire – a four square kilometre site that hopes in time to provide power for 20,000 homes. This was the world's first floating wind farm, Hywind Scotland. A combination of commercially off-the-shelf components and new patents developed and owned by parent company Equinor, the six Hywind Scotland units proudly stand 254 metres high, two and half times the height of London's Big Ben clock tower. The floating wind turbine's rotor diameter is 154 metres and its vertical tubular structure floats upright like a (much smaller) spar buoy thanks to its ballast design. Water depths on the site vary between 95 and 129 metres, with the average wind speed known to gust at around 10 metres per second. Because it floats, of course, it enables operators to push further out to sea, into much deeper waters – to take advantage of faster gusts of wind, to move out of sight of shore-side residents and to minimize the risk of conflicts with other sea activities (shipping, fishing and leisure activities).

Getting the power back to land is obviously still an issue, of course, and Hywind Scotland is served by an export cable 30 kilometres in length to enable safe energy transfer back to shore. As much as 80 per cent of the of the total potential for offshore wind power is estimated to be in deep waters, so the floating wind innovation is exactly what is needed. While Hywind Scotland will generate 30 megawatts, the parent company insists that it is just an initial trial that could eventually lead to commercial operations up to ten times its size. The company states that it is assessing further potential in West Coast USA, Japan, France and Ireland.

Hywind is not alone in its thinking. In March 2019, a new floating wind farm project off the west coast of Ireland was announced, with a price tag of €31 million, led by the European Marine Energy Centre, in partnership with Sustainable Energy Authority of Ireland and engineering company Saipem. A new floating wind turbine site will be developed near Belmullet, County Mayo, by 2022, supported by a grant from the European Union's Interreg North West Europe programme. The seas off this coast are known to have some of the strongest wind resources in the world, so it should come as no surprise that the area is drawing such investment.

While, as we see, the UK and Ireland are leading the way in offshore wind, other European markets also look promising. France, Spain and Portugal have been identified as prime floating wind locations as they all have the key criteria of large and deep territorial waters, significant wind patterns, large populations and intensive, power-hungry industrial activities near the coastline.

Maintenance from afar

Technological innovation does not stop once offshore energy systems are installed and operational, however. As with most technology sectors explored in this book, the offshore renewables sector is embracing two of the more prevalent cross-cutting technologies – autonomous systems and satellite-enabled services. For the latter, opportunities include using satellite-derived bathymetry to map water depths and seabed dynamics when planning new sites close to shore; quantifying metocean conditions such as winds, waves and sea surface height via satellite-mounted altimeters and radar sensors; and providing ongoing situational awareness in offshore renewable energy locations using high-quality Synthetic Aperture Radar imagery.

For autonomous and unmanned systems, one of the most promising areas of crossover in the offshore renewables market is emerging as providing services for operation and maintenance activities. These essential tasks typically represent a large part of the total costs of offshore wind farms (between 25 and 30 per cent of the total lifecycle costs) (Miedema, 2012). In keeping with the growth of the sector as a whole, the global wind operations and maintenance market is projected to grow from just over $13.7 billion in 2016 to around $27.4 billion by 2025, representing a compound annual growth rate of 8 per cent, according to market research company GlobalData (McCue, 2017). Reflecting its pre-eminence in other areas, the UK's share of that market is expected to increase from 5.3 per cent in 2016 to 7.1 per cent in 2025.

Initiated in 2008, the Carbon Trust's Offshore Wind Accelerator is a collaborative research and development programme that involves nine offshore wind developers that jointly account for 76 per cent of Europe's installed capacity. The Accelerator released an underwater inspection methods research study in 2017 that underlined the growing need for innovation in the offshore wind subsea inspection sector. As of January 2017, there were over 3,500 turbines installed in Europe, representing a cumulative total of over 12.5 gigawatts installed capacity. Over 80 per cent of these turbines are built using monopole structures, which have grouted joints where two steel tubes are connected. With European offshore wind capacity predicted to almost triple over the next decade, many more monopile structures will be in situ. All of them need to be monitored. Current estimates show that around 35–40 per cent of the monopile fleet (the majority of the pre-2012 structures) have potentially been affected by issues relating to their grouted joints. Corrosion caused by the harsh marine environment is not uncommon. Additionally, many of the structures built after 2012 will require performance monitoring as they rely on newer design features such as 'jackets', which are constructed using welded nodes, and will require regular inspection when deployed in the field.

Industry standards require that a sample of offshore wind subsea structures within each farm (typically 5 to 15 per cent) are subjected to periodic inspection. This can be carried out at varying intervals, depending on the standards used, but would typically occur every four to five years per wind farm. If defects are observed, such as witnessing greater levels of corrosion than expected, then the inspection programme can call for wider or more frequent inspections, which can range up to all assets in the farm, in the case of a serious generic problem being identified.

Crucially, the accessibility of an offshore wind turbine by ship is largely determined by the height of the waves, which are obviously not always predictable. Weather situations with a significant wave height of more than 1.5 metres are designated as 'weather days', and in such situations the transfer of service personnel from a workboat to the access ladder of the offshore wind turbine is too risky. Even using special access systems that can compensate for the vessel's movements, accessibility is still limited to 2 metres significant wave height.

The demands of offshore wind inspection requirements are therefore perfectly aligned to the emerging potential of autonomous and unmanned systems. Such systems have been specifically developed both to reduce necessary costs and enhance health and safety.

Unmanned and autonomous service offers are being developed to enable offshore wind customers to undertake underwater structure inspection, eg following severe weather incidents or for the kind of routine scheduled inspections set out above, and to undertake seafloor surveys (eg for cabling installation and maintenance) and again for change detection analysis (eg following severe weather incidents or for routine scheduled inspections).

By not requiring the deployment of more expensive traditional subsea equipment (and expensive support boats and additional infrastructure), unmanned systems can support the growing market with more affordable services. Additionally, by not requiring human intervention underwater, health and safety can be significantly enhanced. The delivery of maintenance services can also be regulated more, as there could correspondingly be less need for weather days.

If a reminder of the importance of the health and safety consideration were needed, it came in November 2018 with the tragic loss at sea of an offshore wind contractor in the North Sea. The unfortunate seafarer's vessel, supply ship *Standard Supporter*, was not even taking part in active operations supporting the installation of new monopile foundations at the 200 megawatts Trianel Windpark Borkum West projects at the time of the incident. Even though the ship was only on standby due to adverse weather, a man still went overboard at 10 past 1 in the afternoon and could not be found by the combined efforts of rescue ships and helicopters scrambled from both Germany and the Netherlands.

Continuous operation

UK company Modus Seabed Intervention is one provider of autonomous systems trying to de-risk offshore wind operation using autonomous and unmanned systems. The company deploys its unmanned underwater vehicle systems in a variety of ways, mainly to support survey stabilization and protection (marine trenching), drilling support, construction support and inspection, repair and maintenance. The range of technologies used to undertake those activities include 'trenching class' remotely operated vehicles (ROVs), heavy duty work and inspection class ROVs, and autonomous underwater vehicles (AUVs), whose differentiation underlines the growing maturity of the unmanned vessels sector.

In 2018, working with bespoke offshore equipment manufacturer Osbit Ltd and the Offshore Renewable Energy Catapult, Modus began the trial of an innovative AUV docking station at the National Renewable Energy Centre off the coast of Blyth, Northumberland. The project (Autonomous Vehicle for the Inspection of Offshore Wind Farm Subsea Infrastructure/AVISIoN) was supported by funding from Innovate UK, the Government-supported UK innovation agency. The new approach will enable automated vehicle recharging and the remote uploading of any data acquired during the latest mission. Conversely, the docked AUVs can download new mission commands to set them on their way again.

The development of the capacity for Modus' innovative AUVs to remain onsite offshore, without the need for an additional support boat, provides new levels of cost saving.

Using AUVS in this way was estimated by the consortium to enable wind farm operators to make cost savings of £1.6 million per annum. That figure represents 0.8 per cent of their levelized cost of energy, a measurement system that calculates the total cost of building and operating a power plant over its planned lifetime. Cost efficiencies at that rate would mean that such an innovation could save the European offshore wind sector as much as £1.1 billion, if costed over a 25-year period and factored across all installed capacity. In the earlier research and development phases of autonomous systems, where risk management issues were still being worked through, this was a criticism that we heard often from operators and potential customers who were being asked to contribute or take part in trials: what value can autonomous systems really provide if they still need to be accompanied by additional support boats? While it is important to have additional human eyes overseeing operations during test phases, the sector is maturing beyond that.

'On land' at sea

For those interested in innovative technology, there is regular enjoyment to be derived from the videos emanating from innovative robotics companies such as Boston Dynamics, who are developing a range of innovative robotics systems for a variety of markets. Publicity videos celebrating the latest developments in their two-legged and four-legged systems can enthral, with the footage of one of their 'robot dogs' patiently ignoring a human trying to stop it opening a door by itself to escape a room, attracting much attention and comment in the media.

While most of the unmanned, remotely operated and autonomous systems featured in this book operate under the sea, on the water's surface or in the air over expanses of ocean, offshore renewable platforms provide an additional opportunity for another category of system. Swiss company ANYbiotics recently released a video of what they claim to be the world's first autonomous offshore robot. Their four-legged robotic creation, called 'ANYmal', was shown in November 2018 happily and efficiently trotting around an offshore wind facility in the North Sea. Multiple sensors and an autonomous navigation system enable ANYmal to undertake routine inspections to monitor machine operations, obtain read-outs of sensory equipment and detect thermal hot spots and oil or water leakage. As with the Modus Subsea AVISIoN project, this again replaces inspection tasks usually carried out by humans, thus reducing the need for staff offshore.

Important health and safety aspects aside, as well as overlooking the many and varied environmental benefits available with offshore and other forms of renewable energy, the further development of unmanned and autonomous systems for the offshore renewable sector will help to strengthen the narrative around the thing that perhaps matters most: cost. In an interview for the *Jakarta Post* in May 2018 while on a renewables fact-finding mission to Scandinavia, Indonesia's Energy and Mineral Resources Minister Ignasius Jonan commented that renewable energy must be affordable, or no impact will be possible (Patria, 2018). This is obviously true in every part of the world, and a repeated theme in every Blue Economy technology sector. While the global warming and health benefits of energy not involving fossil fuel will remain a constant draw, any sector receives a boost if it has a strong and irrefutable economic argument at its centre. This has undoubtedly driven the rate of investment in offshore renewable energy in the past few years, and it is encouraging that the cost effectiveness story continues to strengthen as the sector continues to mature and prove its emerging innovations.

In October 2018, for example, energy industry body WindEurope published a European policy blueprint that, if enacted, it claimed would enable offshore floating wind power costs to fall to reach

€40–60/megawatt hour by 2030, down from a starting point at the time of the report's release of €180–200/megawatt hour (WindEurope, 2018). Recommendations that could help drive such ambitious cost effectiveness in the sector included national governments including their ambitions for floating capacity in their National Energy and Climate Plans to 2030, strengthening access to low-cost financing for innovative floating wind projects, and increasing research and innovation funding that focused on projects that would lead to greater cost competitiveness within the emerging industry.

At the time of the report's launch, WindEurope reported that European companies led 75 per cent of the 50-plus floating offshore wind projects in development across the globe, so further investment and policy interventions seem sensible to enhance commercial and environmental advantage still further.

Potential downsides

Despite all of the many cross-cutting positives listed thus far, it would be unusual if there were not some concerns to consider within the broad sweep of growth. The Nature and Biodiversity Conservation Union, a Berlin-based conservation charity, has criticized the negative effects of emerging offshore wind projects in the German North Sea on the native sea bird population. Despite several published reports presenting opposing evidence, the charity claimed in an official complaint to the European Union in March 2019 that the Butendiek offshore wind farm and other wind farms were causing 'massive environmental damage', especially on the Eastern German Bight European bird sanctuary. The habits of certain species of bird were adversely affected by the installation of offshore capacity, according to the group.

Other potential impacts noted in academic studies include underwater noise pollution (especially related to pile driving and other construction efforts) and the danger to marine mammals, sea turtles and fish of collision with and disturbance from vessel movements and other installation activities. Such activities, and the eventual electromagnetic field activity emitted by cables transmitting the produced electricity, could lead to displacement activities or loss of health of marine life. Of course, collision with moving turbine blades is also a risk for low-flying birds. While understanding of such risks needs to be developed further, they will already have attracted the attention of governments and regulators, who will be continually updating the levels of evidence within the environmental risk impact studies that will be required to unlock construction permits and licences.

Cross-cutting value

It would perhaps be short-sighted to just look narrowly at the price of the energy produced as the sum total of value of the sector. Such is the emerging scale and promise of innovative offshore renewable technologies that the economic benefits have much broader reach beyond their simple (if impressive) ability to reduce the cost of energy.

The positive impact on employment, for example, will be considerable if the industry grows as expected. The UK Government expects the number of 'green collar' jobs in the offshore wind industry will rise to 27,000 by 2030. This means that the sector workforce will more than triple the current 7,200 jobs over the period. The Government's Offshore Wind Sector deal also hopes to more than double by 2030 the number of women entering the industry to at least 33 per cent, perhaps even reaching a stretch target of as much as 40 per cent.

Similar green collar job predictions are being made in potential offshore locations across the globe. In September 2018, for example, a report in South Carolina suggested that the US state could benefit from an average increase of 847 offshore wind-related jobs a year all the way through to 2035, spread across all related disciplines of development, installation, operation and component manufacturing.

Onshore sites closest to the offshore location are likely to benefit most from all forms of offshore renewables growth. That can cause challenges as well as opportunities. In October 2018, the

Maryland-based non-profit organization Business Network for Offshore Wind noted that US ports sometimes struggle to keep up with the fast pace of offshore developments. The 'flash to bang' of offshore wind projects once they have negotiated the necessary regulatory approvals can be short, creating pressure on shoreside infrastructure and delivery capacity. This potential bottleneck will surely minimize, though, once the offshore market starts making more urgent strides in the US, and development and delivery patterns become more well known and more easily integrated into ongoing operations.

To underline that, in the more mature European market, trade body WindEurope is taking a proactive stance on the port infrastructure issue. In October 2018 it released a new report emphasizing that up-front investment in specialist port infrastructure of between €0.5 and €1 billion could help to cut offshore wind sector costs by 5.3 per cent. When one considers that more than 10,000 wind turbines are expected to be operational in European waters by 2030, the extra load on ports – to support surveying, installation, operations and maintenance, decommissioning – starts to look extremely urgent.

Looking at cross-sector value, it is especially encouraging that the more traditional oil and gas industry is not just viewing the offshore renewables industry as an irritating disruptor of its own revenue streams but is working to identify the potential of collaboration for mutual benefit. For example, the UK Oil and Gas Authority initiated a £900,000 project in March 2019 to investigate the transition and synergy potential of offshore wind. Areas for exploration included the possibility of powering offshore oil and gas platforms from renewable sources, and how legacy oil and gas infrastructure can be used in emerging renewable fields.

Conclusion

Many Blue Economy sectors have a reach far beyond the seas and oceans that are their main focal point. The offshore renewable energy sector is one such sector, rooted in the power of the ocean but with a huge range of benefits in many tangential industries and domains.

With the technology becoming proven to ever-more reassuring levels, a further acceleration of capacity is predicted across the board. Not every sector will be able to forge ahead as quickly as the offshore renewables sector, but it would be useful for more immature markets to look at the progress achieved in offshore renewables. They might learn from how the development of the technology has been enabled through business, economic and environmental narratives that have unlocked significant critical investment, from both state funding and private investors enticed by growth prospects and potential return on investment.

This also throws a light on the increasing levels of benefit that can come from several progressive waves of innovation that build on each other – from onshore wind to offshore wind to floating offshore wind, with innovations in unmanned and autonomous systems, training, data transfer and analytics all being pulled along for mutual benefit.

The offshore renewable energy sector is, then, in some ways the best of the emerging Blue Economy technology in action – with strong commercial drivers underpinning ambitious new approaches to long-held issues facing humanity.

But the battle is not won yet, and additional pressure needs to be brought to bear to ensure that the current platform for growth does indeed yield the greatest possible returns. It seems as though this will be realized. In the UK, for example, a number of offshore energy partners are collaborating as part of a body called the Floating Wind Action Group, which was originally set up through the industry trade body Renewable UK. Collectively, they are pushing to keep the momentum going in this most innovative of sectors, encouraging further government investment and support.

Such renewed energy and commitment are much needed and welcomed, not least so that these new technologies can begin to make a noticeable dent in the world's reliance on fossil fuels, which casts such a cloud over so many areas of life as we know it.

References

Awea.org (nd) America's new ocean energy resource. www.awea.org/policy-and-issues/u-s-offshore-wind (archived at https://perma.cc/W5YG-BB67)

Cdn.ymaws.com (2018) *Offshore Wind Industry Prospectus*. https://cdn.ymaws.com/www.renewableuk.com/resource/resmgr/publications/catapult_prospectus_final.pdf (archived at https://perma.cc/9DX4-7Q8H)

Department for Business, Energy and Industrial Strategy (2018) *Digest of UK Energy Statistics* https://assets.publishing.service.gov.uk/government/uploads/system/uploads/attachment_data/file/736148/DUKES_2018.pdf (archived at https://perma.cc/P8XY-MEP3)

Eia.gov (2019) Electricity in the United States: Energy explained, your guide to understanding energy. www.eia.gov/energyexplained/index.php?page=electricity_in_the_united_states (archived at https://perma.cc/FC9D-RSHX)

Iea.org (2018) World energy outlook 2018. www.iea.org/weo2018/ (archived at https://perma.cc/YV8E-UQ7G)

IRENA (2017) *Renewable Energy Prospects: Indonesia, a Remap analysis*. www.irena.org/-/media/Files/IRENA/Agency/Publication/2017/Mar/IRENA_REmap_Indonesia_report_2017.pdf (archived at https://perma.cc/8UXM-ZKRG)

IRENA (2018) *Nurturing Offshore Wind Markets: Good practices for international standardization*. www.irena.org/-/media/Files/IRENA/Agency/Publication/2018/May/_Nurturing_offshore_wind_2018.pdf (archived at https://perma.cc/WV38-P53Z)

McCue, D (2017) Wind report: Wind energy operations and maintenance market set to double by 2025. www.renewableenergymagazine.com/wind/report-wind-energy-operations-and-maintenance-market-20170622 (archived at https://perma.cc/3U3M-CT76)

Miedema, R (2012) Offshore wind energy operations and maintenance analysis, PhD, Amsterdam University of Applied Sciences

Patria, N (2018) Indonesia races against time in renewable energy. www.thejakartapost.com/news/2018/05/30/discourse-indonesia-races-against-time-in-renewable-energy.html (archived at https://perma.cc/UW55-GBPY)

Sugianto, K, Helmi, M, Alifdini, I, Maslukah, L, Saputro, S, Yusuf, M and Endrawati, H (2017) Wave energy reviews in Indonesia, *International Journal of Mechanical Engineering and Technology (IJMET)*, 8, pp 448–59. www.researchgate.net/publication/321421362_Wave_energy_reviews_in_Indonesia (archived at https://perma.cc/H5VA-C39S)

Vaughan, A (2018) UK renewable energy capacity surpasses fossil fuels for first time. www.theguardian.com/environment/2018/nov/06/uk-renewable-energy-capacity-surpasses-fossil-fuels-for-first-time (archived at https://perma.cc/RX7E-3Z3N)

WindEurope (2018) Floating offshore wind energy: A policy blueprint for Europe. windeurope.org/policy/position-papers/floating-offshore-wind-energy-a-policy-blueprint-for-europe/ (archived at https://perma.cc/EXP7-PWFN)

05

The cruise industry

Pushing the digital boundaries of the customer experience

The BOLT™ is an exciting new rollercoaster designed to reach speeds of nearly 40 miles per hour over 800 feet of track. Nothing too innovative there, you might think, until you realize that thrill-seekers riding BOLT will do so *while at sea*.

Carnival Cruise Line announced in December 2018 that this new entertainment attraction would be a main feature of their new XL-class *Mardi Gras*, which is slated to undertake its maiden voyage in 2020 from a new terminal at Port Canaveral, Florida. BOLT – built by Munich-based Maurer Rides – features a motorcycle-style car that will propel four riders at a time up to 187 feet above sea level.

This feat of engineering is just one of a wide array of recent innovations across the cruise sector, which like many other consumer-facing industries has been investing heavily in top-of-the-range experiences to attract custom.

The casual observer may have missed how the cruise industry has been working hard to shed its previous reputation of being the preserve of the older generation. While the average age of the international cruise passenger is still 46, that is the lowest it has been for two decades, and nearly 41 per cent of cruise passengers in 2017 were under the age of 34.

Michelin-starred chefs, 'name' interior designers and expensive art collections abound; iPads are provided to control temperature and light settings easily; cabaret has been superseded by the promise of 'immersive performance art'.

When Celebrity Cruises launched their new ship *Edge*, the ship's 'godmother' – the figure who traditionally blesses and names the vessel – was no less than Nobel Peace Prize winner Malala Yousafzai (Bora, 2018). Malala captured the world's attention in 2012 when she was shot in the head by the Taliban during an assassination attempt in her native Pakistan, in retaliation for her activism promoting women's educational rights. Her continued bravery and principled stance as she recovered from a critical condition earned her a global army of followers and well-wishers, and her values have been coveted by many NGOs and commercial entities – including, now, a leading cruise company.

According to Lisa Lutoff-Perlo, President and Chief Executive Officer of Celebrity Cruises, *Edge* 'needed a godmother who would be even more heroic and transformational than the ship itself'. Malala's 'celebrity' endorsement aligned neatly to the corporate task of tackling well-known gender equality targets within the industry, as Celebrity Cruises aims to ensure that 30 per cent of the crew on board *Edge* is female (the industry average languishes at 17–18 per cent).

Such forward thinking is mirrored in onboard facilities, perhaps most obviously signalled in the ship's theatre. Its four stage areas boast three enormous moving projection screens; 16 state-of-the-art video mapping laser projectors that allow larger-than-life visualizations to be displayed on practically any surface; a rotating platform; 10 synchronized panoramic projection screens; and, finally, rotating spiral staircases. Over a hundred and fifty speakers ensure that the audio matches the visual.

Such high-end aspirations and investments should come as no surprise when the market statistics of the cruise industry are taken into consideration. Maritime and coastal tourism is big business; it accounts for 26 per cent of the ocean economy, making it the joint second largest ocean industry (maritime equipment also commands a 26 per cent share). Only oil and gas (at 34 per cent) is of higher economic value in the ocean domain.

Within the maritime and coastal tourism sector, the cruise market has been undergoing steady growth in recent years, with that trend very much expected to continue. The total worldwide ocean cruise industry was valued at about £35.8 billion in 2018, a 4.6 per cent increase over the previous year (Cruisemarketwatch.com, 2019). An estimated 26 million passengers were carried in 2018, which was a 3.3 per cent increase over 2017.

Further industry growth projections are compelling. According to the 2018–19 *Cruise Industry* News *Annual Report*, the global cruise fleet is expected to increase by an additional 86 ships (to reach a total of 472 vessels) over the next decade, representing eye-catching growth of 22 per cent over the period (Mathisen, 2018). As many of these new ships are larger than those being replaced, annual cruise passenger capacity will subsequently grow to reach 39.6 million by 2027, which equates to a staggering 48 per cent increase over the next 10 years.

The growth projections for China alone are revealing, with the country's Ministry of Transport setting out a vision in September 2018 aim to increase the number of cruise passengers to 14 million by 2035, from a quoted total of 2.43 million in 2017. The number of domestic cruise passengers has increased by more than 40 per cent annually since 2006, but even that impressive growth is just the beginning, with the relaxing of a policy allowing foreign tourists from cruise ships to stay in China for up to 15 days without visas – provided they leave the country on the same cruise ships – being one of a number of stated enabling mechanisms.

Such cross-cutting growth will obviously enhance the industry's bottom line considerably, with gross revenue (based on current ticket rates and onboard spending) elevating to £46.3 billion ($59 billion) in 2027, an increase in the region of 50 per cent.

With enhanced potential revenue comes fiercer competition, and the larger industry players are aligning their growth to the promise of a high-tech future that will lead to transformative environmental and business performance. The industry is clearly taking this commitment seriously, with existing and planned technological innovations encompassing everything from smooth customer experience to enhanced energy efficiency, comprehensive recycling and all manner of potential improvements driven by high-powered data analytics.

Greener and cleaner

In keeping with the need to be able to rely on green credentials as a selling point, it is no surprise that – as with all ocean-going industries – minimizing harmful environmental impacts attracts great attention in the world of high-end cruises. And this commitment covers everything from initial ship design to a range of innovative corporate social responsibility programmes.

One notable environmental milestone came in December 2018, when AIDA Cruises took delivery of the world's first cruise ship powered by liquefied natural gas (LNG), currently the most environmentally friendly and lowest-emission fossil fuel. The design of the 180,000 ton *AIDAnova*, designed and built by German firm Meyer Werft and able to accommodate 6,600 passengers, therefore removes the need to use heavy fuel oil.

LNG is natural gas that has been cooled to reach a temperature of minus 162 degrees Celsius (minus 260 Fahrenheit). Emitting about 25 per cent less carbon dioxide than conventional fuels, LNG also contains 85 per cent less nitrogen oxide, only traces of sulphur, and 99 per cent less particulates – exposure to which has been linked to cancer. All of this certainly helps a cruise company to celebrate publicly its commitment to environmental protection, ever more important in the burgeoning world of eco-travel.

A further 18 new build LNG-enabled cruise ships are also on order across the industry (with estimates of 25 more LNG-powered ships in operation by 2025). While this committed order book only represents approximately 25 per cent of new build cruise ship capacity, such green investment is welcome, even though industry analysts are keen to point out that such initiatives also bring with them further investment needs that include revolutionizing the supply chain so LNG is available in large quantities, on multiple cruise routes across the globe. Arranging LNG training for relevant cruise industry staff has also been highlighted as a major endeavour necessary to facilitate significant change.

The move towards LNG is not just driven by potential efficiencies and to provide the opportunity for good customer-facing public relations, however. All cruise companies are very much aware of the shifting policy agenda, particularly the elements that will increasingly affect emissions within the industry. The Norwegian Government's climate strategy for 2030, for example, established the path for additional regulations specifically to target emissions from cruise ships and other shipping in tourist areas, so that they are mandated to phase in solutions that will lead to 'low and zero emissions' (Stortinget, 2018). This commitment is even more forcibly underlined within the country's World Heritage Fjord areas, requirements for which demand entirely zero-free emissions four years earlier in 2026.

The cruise industry's collaborative body, the Cruise Lines International Association, has been driving engagement to address these policy demands. It announced in December 2018 that – as an industry and taking 2008 statistics as a baseline – it was committing to reduce the rate of the global fleet's carbon emissions by 40 per cent by 2030 (Pai, 2018).

Coming in from the cold

LNG is not the only fuel innovation in play, as cruise companies have a broader menu of options to choose from, depending on the environments within which they operate. Amongst other market offers, Hamburg-based operator Hapag-Lloyd run specialist 'cruise expeditions' to the Artic and Antarctica. As these areas are deemed particularly vulnerable, Hapag has been utilizing marine gas oil (MGO) – which has a sulphur content of only 0.1 per cent – when visiting them. The company announced in January 2019 that it was doubling this commitment by extending the use of MGO to its entire fleet of expedition vessels by July 2020.

Indeed, the expedition cruise sector as a whole is proving to be a useful test bed for such energy innovation. This niche sub sector is also undergoing rapid growth, with 28 new expedition ships (specifically built to transport passengers to harsher environments) set to be operational by 2022. With the fleet of ships undertaking polar tourism set to double in the next two years, the collaborative work of industry bodies such as the Association of Arctic Expedition Cruise Operators and the International Association of Antarctica Tour Operators is essential to ensure that such environments remain pristine – free from the risk of oil spill or degradation through repeated exposure to carbon emissions.

Other companies operating in the polar regions – such as French luxury cruise line Ponant – are investing in dual-fuel expedition options, where they utilize low-sulphur fuel oil in order to transit between the poles, switching to LNG when entering sensitive areas.

Further promise in this market comes with the development of the world's first hybrid cruise ship, which is undergoing sea trials in 2019. The MS *Roald Amundsen* – owned by Norwegian specialist operator Hurtigruten – has been built to take up to 530 passengers to harsh polar environments and can undertake (limited) periods of entirely electric operation.

This follows on the heels of another much-discussed energy innovation from Hurtigruten, who announced in November 2018 that from by 2021 six of its 17 ships would be powered by liquefied biogas – fuel made from dead fish and decomposed organic waste – at a cost to the company of $826 million over three years.

Back to the future

In April 2018, Viking Line's cruise ship the *Viking Grace* underlined the fact that innovation can look to the past in order to embrace the challenges of the future. Not content with already being powered by LNG, the passenger ship undertook its maiden voyage with additional assistance from wind power.

The new Norsepower rotor sails fitted on board the *Viking Grace* are 24 metres high with a diameter of 4 metres. They are projected to enable an LNG saving of 300 tonnes a year (equating to a 10 per cent annual reduction in fuel required) and a reduction in carbon dioxide emissions of 900 tonnes a year, all by producing additional thrust from the most traditional of ocean-going vessel power. Built of lightweight composite materials, each side of the rotating surface is designed to attract different pressures of wind, creating a draw towards the area where the pressure is lower. The mechanized spinning sail therefore produces a thrust force that reduces the power required from the ship's main propellers. The system is fully automated, designed to autonomously sense whenever the wind is strong enough to deliver fuel savings, at which point the rotor will start generating power on its own.

Fascinatingly, the technology concept within this context itself is nearly 100 years old, but faced its first death knell as a result of another wave of innovation. The original Flettner rotor (named after engineer Anton Flettner) originally showed great promise in the early 20th century, demonstrating impressive efficiency. However, it was soon superseded by the gains offered by much cheaper fuel, which soon took prominence and condemned the Flettner rotor to – until recently – a footnote in innovation history.

In the 1920s, no less a figure than Albert Einstein highlighted the sometimes slow progress of technology development and adoption. He criticized how long it had taken this specific form of wind power to start to come to market when he said, 'The scientific basis for Flettner's invention is actually already about 200 years old.' Its makers must be hoping that the product has better market longevity this time!

Floating on the digital revolution

Ship design and emissions compliance aside, emerging plans for technology-driven improvements in cruise customer experience are perhaps the most conspicuous area of growth and impact and have thus been attracting much press attention and debate in recent years.

As a potential tourism destination in their own right, cruise ships have to compete against land-based offers, but the cruise industry also appears to be thinking beyond simply trying to replicate hotel offers, to attempt to rival the rich consumer experience that existing or potential customers are becoming used to in their daily lives. So, what kind of experience does the tech-savvy consumer demand these days?

The progression in the past several years of more detailed understanding of user-friendly online and digital experiences created new challenges and opportunities for customer-facing businesses, which in turn were supercharged with the rapid adoption of the smartphone in advanced economies.

Mobile commerce has created significant growth opportunities for most business-to-consumer industries, offshoots of which are a heightening of customer expectations and a higher bar in place for businesses who want to impress in order to secure higher market share.

Back on land, for example, digital technology has been ramping up significantly in the food and beverage sector within the past few years. For one small example, look to the Public House gastropub in Chicago. Far from troubling customers with the burden of physically queuing up to purchase drinks, each table is provided with its own digitally tracked beer on tap. Customers can simply top up their beer glasses themselves (choosing from a selection of brews), tracking how much they have consumed, and how much they owe, via a tablet PC. Aside from the seamless experience, removing the need to queue also increases actual drinking time, so everybody wins.

Taking that up a level, leading retail companies such as Westfield now see their future flagship outlets less as 'shopping malls' and more as 'hyper-connected micro-cities' (Hendriksz, 2018), with digital engagement – and data-driven insights – central to the offer. While their vision – cast forward to 2028 – pledges eye scanners aiding personalization and walkways informed by artificial intelligence, several underpinning technology elements are not actually that far away. More advanced shopping malls have already been experimenting with face recognition software so that sensor-laden digital signage can capture demographic data (Ayonix.com, 2019), though it should be noted that the use of such technology has been criticized by civil liberties advocates where specific consent has not been obtained. In one example in Canada in the summer of 2018, a browser window accidentally left open on a digital display screen revealed that a camera was tracking the approximate age and gender of shoppers using the mall's digital directory (Rieger, 2018). Representatives of the technology provider responded that they were neither capturing nor storing the imagery, so data protection safeguards did not apply.

Similarly, consumers have now become used to accessing high-powered Wi-Fi as they peruse the shops in malls and airports, even if they may not yet be as aware that the Wi-Fi masts through which they connect are also tracking their footfall to identify shopping patterns.

As the chief information officer of Westfield commented in 2017: 'We are increasingly trying to understand real-time dynamic traffic flows' (Pudwell, 2017). Accessing geo-located data can help to identify shopping trends that then allows the malls' customers – the retailers – to understand better where they sit within the 'pecking order' of visits, and how they might better target their offers to specific groups of shoppers.

Whatever the merits of these specific cases – and the data privacy debate will run and run – they underline the fact that what some consider to be the 'future' is very much here already, with digital technology both supporting enhanced user experiences and providing companies with levels of data not previously available.

One of the most compelling examples of the potential power, and direct economic value, of personalized data within a physical customer environment (as opposed to an entirely digital/online offer) comes from the Dallas Museum of Modern Art.

As far back as 2014, the museum took a head-turning step: they made admissions and memberships completely free, if patrons were willing to share some personal data, even if that was just their name and email address. This was an initiative with a significant price tag; in pledging to make memberships free, the museum was in effect immediately offering up the $1.2 million it received from memberships from its bottom line, which equated to some 5 per cent of its annual budget.

However, this was very much considered as a valid investment rather than an eye-catching 'freebie'. When announcing the initiative, Director Maxwell Anderson said: 'We're trying to incentivize people to represent what they're doing, where they're going, and how they're spending their time' (Tozzi, 2014). This was a fundamental and very public change in approach to the perceived value of data to the organization.

The result? Membership tripled in a year. Through associated initiatives, museum staff were also able to track which galleries were most popular, which members were repeat visitors, and what low-income areas were being served (or not served) by the museum most. When considering revenue streams, it was immediately clear to the museum's management that this type of data was critical for grants and fundraising.

Subsequently, free access to the museum increased visitors – who spent money in the café, gift shop, and on special paid programmes. In addition, the museum has been able to use the demographic data from the two million records it accessed as part of the initiative to raise $5 million in grants to support its work. Management thus more than quadrupled the amount of revenue that had previously been paid for admission and memberships.

Getting personal

It is this broader backdrop of providing tech-savvy consumers with the kind of digital experience that they increasingly demand, and a growing understanding of the value of personalized, often geo-located customer data, that has driven the larger cruise ship companies to invest in digital innovation. One of the leading expressions of this drive is the OceanMedallion™, developed and launched by the Carnival Corporation and currently offered on selected Princess Cruise lines, now known as the Medallion Class. The longer-term vision is to integrate the technology on over 100 ships servicing 147 destinations.

In some ways standing in the shoulders of Disney's Magic Band (as they were both designed by the same man), which introduced similar technology to the theme park industry in the shape of a digital wristband, the OceanMedallion™ is a small disc that can be worn around the neck or stored in a clip or a bracelet. It is packed with personalized digital information intended to make the customer experience on board much smoother and more integrated.

Powered by the latest Bluetooth low-energy and near field communication technologies, doors unlock automatically as you walk up to them; digital payments are possible at a wave; personal information can be instantly transferred to speed up the reservation process. The Medallion is even sent to the passenger's home beforehand – with their name, ship name and departure and return dates etched into it by laser – so that their embarkation and disembarkation experience is as smooth as possible. No need for tickets or boarding passes.

When purchasing beverages or food on board, as the Medallion is geo-located, passengers can order drinks (via their connected smartphone or tablet running Carnival's Ocean Compass app) wherever they are and be found easily by crew members to be served. You then don't have to fumble to find your credit or debit card or wipe sun cream off your hands in order to sign the receipt to charge the bill to your room.

Moreover, rather than just focusing on such convenience, the system uses cutting edge data analytics and machine learning to push towards better and more intuitive customer service. The system keeps track of your past orders so that staff can customize your meals – or be ready to pour your favourite drink just as the landlord of your long-time local would – without you even having to ask.

No more cumbersome texting of friends and family members who are trying to meet up with you, as the Medallion allows you to share your location with them automatically. Don't know how to get where they are? The Medallion will guide you.

These important geo-location features are made possible by an impressive network of some 7,000 onboard location sensors, wired into the ceilings of every deck, that in effect act as a bespoke GPS system. Carnival calls this embedded network the xIOS, or the Experience Innovation Operating System. Additionally, participating ships need to be fitted out with miles of cable, hundreds of readers and more than 100 interactive portals.

The Medallion itself does not obviously look like a digital product: one of the major design challenges came about because the product leadership team were keen for it to properly feel like a medallion – ie metal, not plastic. Beyond aesthetics, it has no menu, no on/off button and requires no charging. If you don't want to download the app, you can interact with the Medallion system via your in-room television or at special kiosks dotted about the ship.

Thus, the powerful personal Medallion removes the need for tickets, wallets, keys (all hard to carry and locate when in swimwear) and pesky activities like walking and queuing, instead providing digitally driven luxury customer service and relaxation.

Safe and secure

Crucially, of course, Carnival also says that no secure personal information is stored on the Medallion itself, so there's no risk that someone might try to hack it. The device merely acts as the portal to your securely encrypted digital profile. In case the Medallion falls into someone else's hands (by accident or more nefarious means), staff will be able to spot anomalies at the point of purchase as any engagement will display your onboard security photo.

Carnival first announced its intention to launch the technology at the Consumer Electronics Show – the consumer tech industry's annual flagship event – at the Las Vegas Convention Centre in January 2017. Speaking at the time, Arnold Donald – who became Carnival Chief Executive Officer in 2013 – asked: 'Who wouldn't want to feel like they had a constant personal concierge with them every second?'

After following what many see as the standard 'hype cycle' where potential has perhaps taken longer than anticipated to match up to promise, the wearables market has been progressing steadily in recent years. The latest smart watches, for example, allow users to track all sorts of detailed health activity, geo-track fitness activity, remember where you've parked your car, stream music and pay for coffee via near-field communication. However, some of that engagement activity relies on the underpinning sensor infrastructure being available within the real world, in most cases the retail environment.

Luca Pronzati, Chief Business Innovation Officer at MSC Cruises, underlined the added challenges of digital integration at sea when he described MSC's new generation of ships as akin to 'smart, connected cities, but with the added complexity of being at sea'. However, there are also specific inherent advantages to the cruise ship environment, and the cruise sector has started to capitalize on the key feature of having a captive audience in a relatively confined space. For many companies, this provides the platform to push the boundaries of digitally driven wraparound customer services, supported by a comprehensive Internet of Things sensor network. The personalization elements embedded in the system – where over time the Medallion will better understand passenger preferences to be able to tailor offers based on their purchase and activity history – are made possible due to Big Data analytics capabilities and complex machine learning.

John Padgett, Chief Experience and Innovation Officer at Carnival Corp, has characterized the embedded personalization as being driven by a 'guest genome' within the disc that utilizes hundreds of algorithms to constantly update in order to create 'experience intelligence' after every interaction. It is the near-real-time nature of this data crunching that may be one of the game-changers. Rather than – as is becoming standard in many customer-facing segments and as utilized so successfully by the Dallas Museum of Art – allowing customer data to be collected and analysed to update user preferences for future personalized marketing activities, or to enhance experiences on a future potential vacation, Carnival uses it to try to make the passenger's very next interaction a better one. As Padgett puts it: 'Our strategy is to study you to make the experience great in real-time.'

Of course, Carnival is not alone in investing in such leading-edge assistive digital technology within the cruise sector. Leading competitors such as MSC Cruises and Royal Caribbean have also highlighted their own billion-dollar investments to underpin digital transformation of the customer experience.

Royal Caribbean partnered with EY to develop its new digital strategy, and – mirroring similarly ambitious statements across the industry – the global consultants commented that this was less about just developing new apps, and more 'an opportunity to rethink how to go to market and how to operate the entire business' (Ey.com, 2018). The delivery of this ambition does indeed involve a broad range of innovations. Robot bartenders (or 'bionic mixologists', as the company calls them) now shake up cocktails for the thirsty passenger. Harnessing the latest biometric technology allows facial recognition software to streamline the check-in and boarding process, early trials of which – conducted with US Customs and Border Protection at Cape Liberty Cruise Port in New Jersey – suggested a 40 per cent reduction in passenger processing time over traditional, manual processes. Passengers are able to track their personal luggage via personalized radio frequency identification tags. Finally, the use of virtual reality technology allows passengers both to enjoy gaming on board and to check out shore excursions in 'try before you buy' virtual reality experiences first before committing to them fully.

Okay, Zoe?

Early in 2019, Geneva-based MSC Cruises announced new details of Zoe, its 'virtual personal cruise assistant', another world first, according to the company. Zoe is a voice-enabled assistant, much like the Amazon Alexa or Google Assistant systems that have been introduced to the domestic technology agenda in the past few years. Launching first on brand new ship *Bellissima*, and subsequently on all new MSC ships, passengers will be able to ask Zoe all manner of questions about their voyage, book excursions, check billing, etc.

As you might expect, Zoe launches with the promise of seven languages (English, French, Italian, Spanish, German, Brazilian Portuguese and Mandarin) and, as with the Ocean Medallion, promises to utilize machine learning in order to keep deepening its understanding of customer preferences.

The technology was developed with Samsung Electronics subsidiary Harman Connected Services and utilizes far field and quad-core processors to provide the fastest possible response times. Natural speech recognition and the ability to convert speech to text (and vice versa) offers passengers the opportunity to access information in a much more user-friendly manner. More conventionally, travellers can also connect their phones, laptops and tablets to Zoe by bluetooth in order to listen to their own choice of music.

Of course, by providing answers in this way to 800 of the most frequently asked onboard questions (with thousands of different variants programmed in for each question), the system relieves pressure on crew, so they can focus on more detailed, specific support where it is needed.

Elsewhere on the MSC fleet, children can pick from a broad menu of technologically-driven entertainment – everything from virtual reality gaming to being able to create digital designs and then print 3D versions (using HP's Sprout technology) to take back to show their parents.

Reaching for the stars

Finally, the cruise industry understands there is one final area that needs significant further investment to keep its tech-reliant passengers happy in the digital age: connectivity. If all other onboard digital experiences and infrastructure are cutting edge, passengers will take an even dimmer view if Wi-Fi speeds are as slow as they have been in a sector that has been grappling with this issue for some considerable time.

Again, this challenge mirrors similar development barriers in other sectors, so there is considerable broader learning to be accessed. American sports stadia have led the way in recent years in grappling with the Wi-Fi challenge, to allow visitors to access services. Anyone who has attended a major sporting event will have experienced the frustration of a communication blackout, as tens of thousands of attendees try to get online via their mobile phones at the same time. This is especially problematic for an industry (much like the cruise industry) keen to attract the younger generation who expect to have continuous access the internet and to be able to share their lives instantly across social media channels.

The US National Football League – and individual American football clubs – tackled this head on, investing in the very latest Wi-Fi infrastructure to allow fans to stream to their hearts' desire. The progress of this is often measured in the data usage experienced at the stadium hosting the Super Bowl. US Bank Stadium, for example, which hosted Super Bowl 52 in 2018, broke data usage records for the fifth year in a row, with fans using 16.31 terabytes of data over Wi-Fi – the equivalent of approximately 40 million selfies. To meet this connectivity need, operators are turning to the twin saviours of enhanced satellite connectivity and robust very small aperture terminal (VSAT) ground station/antenna hardware.

Cruise ships have in recent years provided internet connectivity by accessing a number of defined satellite options. These include Inmarsat's Global Xpress constellation (Ka-band) from geostationary satellites (C-band and Ku-band). Satellite companies such as Luxembourg's SES Networks and IntelSat, however, have caught on to the digital explosion in the cruise industry – and their growing customer needs – and have accordingly been investing in new on-orbit capacity to provide additional throughput in regions of high demand. Crucially, these new constellations are medium-orbit, offering much greater connection strength than the patchier high-orbit options traditionally serving cruise companies.

Back on Earth, such connectivity is only as good as the technology that channels it, which is why the larger cruise companies are also investing heavily in VSAT technology. VSAT underpins all of Carnival's Ocean Medallion capabilities, and the company loudly celebrated how its tri-band antennae allowed it to break the world record for bandwidth at sea in March 2018 when it achieved 2.25 gigabits per second (compared to an industry standard of 10–20 megabits per second) during an event specially tailored for the purpose (Wingrove, 2018). VSAT may now have become a prerequisite for any company wishing to compete for bragging rights regarding onboard connectivity.

Conclusion

Far from being seen in some quarters as the preserve of the middle-aged and elderly, with traditional analogue facilities to match, the global cruise industry is very much aware of the changing technological expectations of its customer base. Despite being at sea, the key industry players are fully aware that they compete with on land entertainment options, the digital experiences those competitors are able to offer with more easily accessible connectivity and the burgeoning expectations of a customer base used to such digital richness.

Those ever-more rigorous expectations stretch to the environmental realm, too, with additional pressure on green credentials being demanded by a more discerning and challenging customer base and the progression of a policy agenda that knows significant action is required to mitigate against potentially harmful actions.

The cruise sector is investing heavily in a broad range of innovative technology options – and in many cases the new ways of thinking that are both driven by and rely on the availability of innovative options – to provide solutions to these challenges, and therefore allow it to meet its own ambitious growth targets. While the end points of some of this investment already look clear, as in many sectors, in time it may be that it is the full utilization of Big Data and machine learning integrations that lead to new paradigms.

As the digital and Big Data innovation experiment beds in across the Carnival fleet in the coming years, to take one example, the next stages of development that this might make possible cannot not yet be fully grasped. The company has stated that it plans to invest 3–5 per cent of capital expenditures each year in research and development projects, so – enhanced customer experience aside – it will be interesting to track how the mass of rich customer data that Carnival now habitually collects may influence the focus of their investments, and potentially even move beyond customer experience to more fundamental shifts in design thinking.

Much as Westfield and other forward-thinking retail sector companies on land are collecting geo-located footfall data to influence strategy, sensor-rich cruise companies – who now collect vast quantities of data as a matter of course, and have the machine learning wherewithal to utilize it – may soon also be presented with new intelligence to guide more lateral strategic thinking, driven by detailed understanding of what customers actually spend their time doing on board, the richness of which could only previously have been witnessed in dreams.

References

Ayonix.com (2019) Solutions. https://ayonix.com/solutions/#commerce (archived at https://perma.cc/HC33-SNBY)

Bora, T (2018) Why is Malala the godmother to this cruise ship? www.cntraveller.in/story/malala-godmother-cruise-ship/ (archived at https://perma.cc/95X6-CR52)

Cruisemarketwatch.com (2019) Market share. https://cruisemarketwatch.com/market-share/ (archived at https://perma.cc/BR3J-GYDW)

Ey.com (2018) How digital transformation opened new channels for growth. www.ey.com/en_gl/digital/how-digital-transformation-opened-new-channels-for-growth (archived at https://perma.cc/LRH3-JZGW)

Hendriksz, V (2018) Westfield unveils the future of retail: 'Destination 2028'. https://fashionunited.uk/news/retail/westfield-unveils-the-future-of-retail-destination-2028/2018060430007 (archived at https://perma.cc/R6W4-FVHQ)

Mathisen, M (2018) Global cruise capacity up 48 percent over next 10 years. www.cruiseindustrynews.com/cruise-news/18769-global-cruise-capacity-up-48-percent-over-next-10-years.html (archived at https://perma.cc/C63W-KM3P)

Pai, L (2018) Cruise industry to cut carbon emissions 40%. www.marinelink.com/news/cruise-industry-cut-carbon-emissions-461051 (archived at https://perma.cc/9L7T-PCNH)

Pudwell, S (2017) Westfield CIO: Data and personalization are key to shopping centre survival. www.silicon.co.uk/data-storage/bigdata/westfield-data-personalization-shopping-215561 (archived at https://perma.cc/BJ5N-7PNV)

Rieger, S (2018) At least two malls are using facial recognition technology to track shoppers' ages and genders without telling. www.cbc.ca/news/canada/calgary/calgary-malls-1.4760964 (archived at https://perma.cc/GW3U-Y4ND)

Stortinget (2018) Climate strategy for 2030: Norwegian transition in European cooperation. www.stortinget.no/no/Saker-og-publikasjoner/Vedtak/Vedtak/Sak/?p=69170 (archived at https://perma.cc/L3NV-BF8R)

Tozzi, J (2014) Dallas Museum of Art trades memberships for data. www.bloomberg.com/news/articles/2014-02-20/dallas-museum-of-art-trades-memberships-for-data (archived at https://perma.cc/BD6V-56ND)

Wingrove, M (2015) Shipping slow to adopt condition based maintenance. www.marinemec.com/news/view,shipping-slow-to-adopt-condition-based-maintenance_41206.htm (archived at https://perma.cc/NUZ6-ZWMZ)

Wingrove, M (2018) Ground-breaking VSAT technology including tri-band antennas helped Carnival drive VSAT to Gbps levels with its MedallionNet connectivity. www.marinemec.com/news/view,carnival-smashes-maritime-bandwidth-record_51077.htm (archived at https://perma.cc/46NE-PZF7)

06

Maritime surveillance

Maintaining eyes on the sea

At precisely 09:28 local time, the fuel ignited in a burst of flame and thick cloud to ease the pristine red and white rocket off the launch pad, and bring forth a hubbub of cheers, high-pitched whistles and enthusiastic applause from the ground crew, excited and relieved in equal measure.

The 39th flight of India's Polar Satellite Launch Vehicle (PSLV), which blasted off from the Satish Dhawan Space Centre, Andhra Pradesh, was a record-breaking endeavour. The headline asset the PSLV was carrying into space was the Cartosat-2D satellite, operated by the Indian Space Research Organization (ISRO) for the Indian Remote Sensing (IRS) programme. The fifth of the Cartosat-2 Earth imaging satellites, Cartosat-2D's payload includes panchromatic and multispectral imaging sensors.

Of greater significance, however, was what else the rocket was carrying into orbit. On the same launch, ISRO itself was also sending a pair of small research satellites (designated as ISRO Nanosatellite 1A and 1B (INS-1A and 1B)) and, remarkably, an additional 101 cubesats. This total of 104 satellites on one launch completely smashed the previous high of 38 from June 2014.

This is the smallsat revolution in action.

Within the space industry, traditional aerospace companies are being aggressively challenged by smaller and – they argue – more agile, responsive and lower-cost commercial companies, often start-ups. This shift is being driven both by a reduction in the costs of operating in space, and a more vigorous demand for satellite data from the market. Nowhere is this trend more evident than in the rapid emergence of small satellites (smallsats), and a range of associated products and services.

Small satellites (generally considered to be anything up to 500 kilograms in mass) operate in low Earth orbit (LEO) 450 to 650 kilometres above the Earth's surface. They take 90–100 minutes to complete one orbit and pass over any geographic location on Earth periodically, usually once a day. Each transit over a location (a 'pass') takes approximately one to twelve minutes, depending on the specific orbit type and altitude.

What is most remarkable about them is their size. The smallest, cubesats, can be held in the palm of your hand. They measure 10×10×11.35 centimetres, with a mass of no more than 1.33 kilograms per unit, yet they can perform an impressive range of tasks when in orbit.

The main rationale for the continued growth of such miniaturized satellites is their ability to reduce associated cost: heavier satellites require larger rockets with greater thrust, and cost more to finance. In contrast, smaller and lighter satellites require smaller and cheaper launch vehicles and – as demonstrated so effectively by ISRO – can be launched in significant multiples. They can also be launched 'piggyback', using excess capacity on larger launch vehicles.

The miniaturization of this technology, then, which in turn is leading to a rapid reduction in costs, is having a transformative effect on the industry, which is said to be moving from 'old space' (where large entities, often government agencies, build, launch and manage a small number of space-based assets) to 'new space', where a much larger range of smaller companies are developing, launching, managing and extending satellite capabilities.

By lowering the barriers to entry in this way, this shift is enabling easier access to the space industry for so many more players who previously could not engage. Smallsats also enable missions that a larger satellite could not accomplish, such as constellations for low data rate communications; using formations to gather data from multiple points; in-orbit inspection of larger satellites; and a broad range of university-related research.

Accordingly, the smallsat manufacturing and launch services market alone is anticipated to grow to a value of £27.8 billion by 2027, based on an estimate of an incredible 6,500 smallsat launches during that 10-year period. This compares to the 2015 baseline of 335 satellite launches across the whole industry (NSR, 2018).

Once they are in orbit, this creates a whole new set of opportunities for new customers to access cost effective, highly accurate global data collected from space, often in real time. And this ability – known as remote sensing – is a major boon for the Blue Economy, which has a massive expanse of ocean that it needs to track and observe.

While market analysts predict that new communications capabilities will drive the largest share of smallsat revenues in years to come, with some of the largest planned constellations addressing this segment, the growth in Earth observation and situational awareness capabilities are also notable.

Northern Sky Research predicts that launch rates for situational awareness initiatives will grow at a 21 per cent compound annual growth rate by the year 2025, and that the proliferation of small satellites will act as a primary driver for the situational awareness market, as LEO constellations offer high revisit rates, significantly lower latency (how long it takes to be able to access that data) and smaller amounts of capital expenditure (Russell, 2017). Northern Sky also suggests that the market for data, value-added services and information products from satellite-based Earth observation will grow to $5.1 billion by 2023, up from $2.1 billion in 2013.

Just focusing on the 2017 ISRO PSLV, the majority of the cubesats it carried were Earth-imaging satellites owned by US company Planet Labs. Additionally, eight of the cubesats it launched (joining Spire Global's Lemur-2 constellation) carried a meteorological payload that collects data related to atmospheric temperature, humidity and pressure. Their payload also included SENSE receivers for the global automatic identification system (AIS), thus enabling them to collect essential tracking data from ships at sea.

These added 'eyes in the sky' could not come at a more critical time for a broad range of ocean-based industries. The European Union helpfully defines maritime surveillance as 'the effective understanding of all activities carried out at sea that could impact the security, safety, economy, or environment of the European Union and its Member States'. The aim of maritime surveillance, then, is to understand, prevent (where applicable) and manage the actions and events that can have an impact, be that positive or negative, on all related activities and sectors.

Security services, shipping companies, NGOs, oceanographers, fishermen, climate change specialists, sailors and many more who traverse have an interest in or rely on the oceans, all have a thirst for maritime surveillance information, and are all eagerly encouraging the increased gathering of that data in the smallsat market.

With reference to security issues, as US Coast Guard Rear Admiral Brian Salerno wrote so eloquently in the Harvard Law School's *National Security Journal* in 2015: 'We cannot hold polluters accountable unless we can match them to their spills; we cannot keep vessels from colliding if we don't know where they are; we can't rescue survivors unless we find them; and we cannot intercept those who would do us harm if they are able to blend in with the millions of recreational boaters who lawfully enjoy our ports and coastal waters' (Wilson, 2015).

In January 2019, the US Coast Guard announced an initiative to help meet that broad range of challenges – a plan to launch two new 60-centimetre squared cubesats (named Yukon and Kodiak) in November 2019, aboard the SpaceX Falcon 9 rocket. Aligned to two new ground stations, the coastguard will be able to monitor and control its satellites directly as they orbit the poles approximately every hour and 40 minutes, between 690 and 1,000 kilometres above the Earth. Yukon and Kodiak will focus on the Arctic, as part of the US Homeland Security's Polar Scout programme, in particular looking to enhance the ability to detect emergency position-indicating radio beacons that have been set off by distressed mariners.

While many coastguard agencies have utilized a wide range of satellite data for many years, these two cubesats were the first ever to be dedicated entirely to a US Coast Guard mission. As we explore elsewhere in this book, the amount of commercial shipping and cruise ship traffic is predicted to increase rapidly in the polar regions in the coming decade, as new technologies enable greater access, and the ice melts due to climate change. Despite this, the Arctic will still obviously remain an extremely challenging, risk-laden environment, so it is encouraging that the smallsat revolution is enabling responsible agencies to keep pace with this change in the development of ocean-focused satellite capabilities.

Similarly, December 2018 saw the launch of what was hailed as 'Africa's most advanced nanosatellite' – the ZACube-2, which 'piggy-backed' the Russian Soyuz Kanopus mission from Vostochny spaceport. The South African Department of Science and Technology announced that the ZACube-2 would enable real-time monitoring of disasters and other emergencies, both natural and man made. In particular, its mission is to use AIS to monitor the movement of ships along the South African coastline.

Weighing just 4 kilograms (roughly the same as an adult domestic cat), the ZACube-2 is another example of a government being able to launch the kind of space-enabled maritime domain awareness initiative that might previously simply not have been possible due to budget and technical constraints.

SAR-struck

While smallsats will, as shown, increasingly bring exciting capacity to the sector, continued investment in more traditional, larger infrastructure is also strengthening maritime surveillance capability.

For instance, in November 2018 the first images were released that had been collected from the new NovaSAR-1 satellite, which had been launched two months previously. Operated by the UK's Surrey Satellite Technology Limited, and supported by a partnership of the UK, Indian and Australian Space Agencies, it is expected to have a lifetime of seven years and focuses on the provision of low-cost satellite imaging.

As its name suggests, NovaSAR-1 provides synthetic aperture radar (SAR) imaging, and the first batch of images highlighted the mission's capabilities and specific areas of focus in the maritime domain. SAR is able to 'see' through clouds, and works at any time, night or day, which provides much greater reliability of coverage. Aligned to the fact that NovaSAR-1 can monitor an area of ocean up to 400 kilometres wide, this is a powerful new tool with many uses.

One set of innovators building services around the use of SAR imagery can be found in a UK company called OceanMind. By employing high-powered machine learning algorithms, OceanMind specialist analysts are able to combine SAR images with AIS activity to pinpoint where potential 'dark vessels' may be attempting to conduct illegal activity at sea.

Since 2004, all passenger ships, vessels over 300 gross tonnage in weight and cargo vessels over 500 gross tonnage have been required by law to have an AIS system fitted on board. An AIS transceiver collects and sends a broad range of data. Information such as its unique identification number (Maritime Mobile Service Identity, or MMSI), navigation status (eg if the vessel is anchored or underway using its engines), speed, position and heading will be transmitted between every two and ten seconds, depending on the vessel's speed, or every three minutes if at anchor. Further details, such as its name, type of ship and cargo being carried, IMO identification number, the type of positioning system it uses and the estimated time of arrival at its intended destination, are subsequently transmitted roughly every six minutes.

What this system provides is a near-real-time picture of the global ocean-going fleet – where vessels are, where they are heading, and, if appropriate, what cargo they are carrying. The Marine Traffic website enables anyone to get a sense of the flow of marine activity, as well as zoom in on individual stretches of water or search for specific vessels.

Such information is obviously of particular interest to regulatory authorities and security services on the look-out for suspicious activity, either as it is happening or to be able to collect evidence when undertaking investigations.

Vessel captains retain the ability to turn off their AIS. When they do so, they are said to go 'dark', off the grid. Previously, this could have provided very useful cover for those attempting to hide their tracks while undertaking nefarious activities at sea. However, the combination of SAR imagery and Big Data applications such as those operated by OceanMind is changing that. By overlaying the visual imagery provided by SAR detections of vessels with known AIS activity in a specific region, it is possible to identify vessels that have turned off their vessel tracking and are 'dark'. Even though the AIS is switched off, the SAR satellites still 'see' the vessels in given areas of water. Having identified vessels, they can then correlate that image with AIS activity in that area at that time and begin to paint a picture of activity and explore why any vessels in a given location may have chosen to turn off its AIS.

There may be valid reasons for turning off a ship's AIS. For example, Article 21 of the system's governing rules state that captains may switch off the AIS if its continual operation has the potential to compromise the safety or security of the ship. It cites the specific case of being 'in sea areas where pirates and armed robbers are known to operate', though it also says that such actions – and the reasons for them – should always be recorded in the ship's logbook. In addition, some fishing captains – operating perfectly legally – have passionately defended their right to turn off their AIS so that competitors are not able to see where they have chosen to operate, in case that gives away any potentially hard-won intelligence of fish abundance.

However, many captains undoubtedly see AIS as a regulatory nuisance, impeding their main goal of conducting illegal activity. As on land, the seas and oceans can host a wide range of criminality, including the smuggling of drugs and other contraband, people trafficking, piracy and intentional marine pollution (eg discharging sewage or oil rather than disposing of it appropriately). As the seas and oceans are so vast, it is impossible to police them adequately with surface vessels alone. However, the ability to utilize satellite data, including SAR imagery, changes the game.

For OceanMind, that is most important in the fisheries domain. They use SAR from all major satellite providers and correlate it against thousands of analysed detections from around the world of known vessel types and ship tracking data. If any differences are noted between the two, this is flagged as a possible indication of potential illegal activity. They are most interested in fishing vessels, so the systems home in on those types, often in specific areas where they are working in partnership with government to detect and prevent illegal fishing.

As well as potentially investigating individual cases, broader strategic opportunities open up through sustained surveillance and analysis, as the resulting mapping of 'hot spots' of potentially illegal activity enables more effective targeting of sea patrols to monitor and interrogate potential transgressors. We explore the capability of the OceanMind system in combatting illegal fishing in greater detail in Chapter 10, but countering illegal fishing is only one option for it and similar satellite-enabled technologies in the Blue Economy.

Illegal fuel smuggling, for example, is a major international problem, its total financial cost estimated to be second only to the global trade in narcotics. Taking one country alone as an example, the Asian Development Bank estimates that the Philippines' foregone revenue due to fuel smuggling amounts to £565 million annually. Some smuggling activities such as this, and crime such as illegal fishing, can be considered to be 'victimless'. However, that is a naive view. Aside from the fact that smuggling is often part of much broader harmful activity within associated criminal networks (eg slavery at sea, threats to and coercion of customs officials, terrorism and other highly syndicated crimes), the loss of additional legitimate financial resource can hit developing economies especially hard. For example, losing the £565 million revenue associated with illegal fuel smuggling deprives the Philippine Government of a critical resource that could pay for any one of the following, every year:

- over 45,000 new classrooms;
- 17 new textbooks for every school-age child in the country;
- the construction of over 375 new 80-bed hospitals;
- the food needs of over 30,000 children for an entire year;
- building construction capacity equivalent to the floor space of 12 Burj Khalifas [at average Philippine construction costs per square metre];
- over 150 new 50-metre offshore multi-mission vessels to help patrol Philippine waters.

So, it is very much in the interests of the Philippines – and all other coastal nations – to stop illegal fuel smuggling, and that fight has, increasingly, been taking place at sea. While coastguard estimates suggest that 80 per cent of fuel smuggling incidents occur directly in the landing/offloading port through, for example, bribery of local officials or adjustment of the cargo manifest during the vessel's passage to its destination, some take place further out to sea. Large tankers full of refined fuel approach the edges of a country's Exclusive Economic Zone and transfer their fuel to local vessels. With their AIS beacons turned off, these vessels avoid paying tax, thus denying the associated government the revenue that it requires to fulfil its mission on behalf of its people. High seas fuel transfers (or 'transhipments') also present a high risk of at-sea pollution, and fuel fraud has the potential to cause harmful auto emissions due to adulterated fuel products in the supply chain.

Dark matters

Another option for the professional fuel smuggler is to bring his or her vessel into a country's waters with its AIS turned off, then deliver its cargo to specified oil refineries, out of sight of the authorities and aided by lack of resource and training, as well, perhaps, as institutional weakness and corrupt local officials, who do not declare such deliveries.

An additional angle to this story relates to the imposition of trade sanctions, such as the oil export ban on Iran announced by the United States in 2018. Iran has been in this place at previous times in its history, and satellite data and machine learning capabilities suggested that some of the threatened exporters had strong ideas about how to circumnavigate the ban.

Tel-Aviv-based maritime risk experts Windward Ltd announced in November 2018 that they had been working with US firm Tanker-Trackers.com and French data analytics company Kayrros to utilize satellite data to track the suspicious movements of Iranian oil tankers. Two months previously, Windward stated, its technology had detected three tankers turning off their AIS while leaving an Iranian port. Days later, it reported, the ships returned to port empty.

This was, Windward said, part of a growing trend as Iranian companies seemingly prepared for the sanctions to start in November 2018. What started as a trickle of vessels turning off their AIS in August of that year turned into a stream by September, with the last two weeks of October, the month before the sanctions were due to start, seeing a 'total blackout'. As Samir Madani, the Chief Executive Officer of TankerTrackers.com commented at the time, they had to verify every single vessel visually using satellite imagery, as no AIS data was coming out of the area at all.

Similarly, in the summer of 2018, western and east Asian intelligence agencies investigating compliance with sanctions imposed on North Korea used satellite data to quantify the amount of fuel smuggling that was taking place in the Yellow Sea, the maritime trade route between Korea and China. The agency exercise counted 148 secret transhipments at sea over just a few short months, activity that involved at least 40 vessels and 130 companies; many of these, it was stated, had Chinese or Russian ties. Collectively, it was estimated, this activity was likely to have seen the transfer of between 800,000 and 1.4 million barrels of fuel. This repeated clandestine business seriously undermined the impact of sanctions on North Korea and thus helped to stabilize its economy.

One example highlighted by the UN involved the *Patriot*, a Russian-flagged tanker. It was seen to switch off its transmitters before linking up with another vessel that was known to have previously undertaken tasks that violated the imposed sanctions. Later, it was noted that the *Patriot*'s draft was changed, which signified that its cargo had been discharged.

This and many other instances underline the growing importance of models that can utilize satellite imagery to cross-validate activity to better understand patterns of behaviour. If ships smuggling fuel (or any form of contraband, for that matter) turn off their AIS, and avoid the chances of having their paperwork checked by offloading their illicit goods at sea and therefore never entering a foreign port, the agencies' 'eyes in space' become critical.

The powerful combination of satellite data and machine learning algorithms make such assessments possible, as millions of data points that might on their own seem innocuous can be combined to deliver a more detailed understanding of a situation. For example, vessel speeds and images of tanker shadow formations (which indicate how high or low it sits in the water) can combine to provide greater accuracy related to likely cargo levels.

Oil on the water

Much of this chapter has focused on the security elements of maritime surveillance, but environmental concerns are also an important element, with many researchers, for example, progressing systems to identify potential oil spills from space. One recent collaborative project in this domain is the Earth and Sea Observation System (EASOS) which used UK Space Agency funding (aligned to overseas development aid funding) to work with Malaysian authorities to explore how satellites could be used to provide an early warning system related to a range of environmental issues, including the identification of oil spills at sea.

Spearheaded by the UK's Satellite Applications Catapult, the project was specifically focused on the Malacca Straits, one of the busiest shipping routes in the world. It is estimated that a ship sails through these waters every 3.6 minutes, collectively carrying £750 billion worth of cargo every year. With such huge traffic evident, the potential of oil pollution caused by operational discharges from oil tank and cargo area cleaning creates an ever-present significant risk to the mangroves and coastal communities on peninsular Malaysia's western coastline.

To tackle the challenge of making sense of unprecedented amounts of marine surveillance data in a manner that can complement existing systems, the EASOS Marine Watch system uses processing algorithms to detect oil slicks using radar satellite data. Where a potential oil slick is identified, this leads to a polygon being overlaid onto various map options. These data visualizations indicate the location of the oil at the time the satellite images were captured. But the system does not stop there. Once a slick is detected, Marine Watch uses weather data and wind forecasts to predict where the slick will travel to over the next three days, to varying degrees of certainty. With limited response resources available, this can help responsible authorities both to improve the detection rate of potential marine pollution events, as they get to see issues that may not have come to their attention through existing means, and to play forward where the potential spills may travel in order to prioritize investigating those that are predicted to make landfall, rather than those that are estimated to float (relatively) harmlessly out to sea. As well as helping in the continual fight against loss of habitat, coastal erosion, species extinction and depletion of fish stocks by preventing ships from pumping their bilges in the Malacca Straits, the system provides the potential to identify where oil slicks may have originated and help intercept offenders.

This convergence of satellite-driven Earth observation data, visual recognition algorithms, machine learning and automated predictive analytics has the potential to revolutionize traditional systems that rely on human eyewitnesses (or, in some cases, static buoys fitted with oil detection sensors) as their primary starting point.

Digital dodging

It is important not to overlook the fact that it is not only the 'good guys' who are aware of and able to utilize satellite and other innovative technologies in related ocean domains. Maritime crime is big business; piracy, illegal fuel smuggling, drug smuggling, human trafficking and illegal fishing all have economic costs that run into the billions of pounds. Sophisticated criminal networks are not just going to sit down and roll over in the face of new techniques and technologies being used by the responsible agencies and increasingly tech-savvy NGOs, non-profits and social enterprises.

The cat-and-mouse game of identifying criminal activity at sea has always involved ever-changing levels of subterfuge. In relation to the illegal fuel transhipments referenced above, vessels are known to adopt fairly crude disguises as they set out on their illicit journeys – from painting fictitious names on their bows to constructing temporary fake infrastructure on deck in an attempt to pass themselves off as a cargo ship rather than a fuel tanker.

But other, more modern techniques are also at play. Some ships have been observed by UN analysts to intentionally broadcast a series of contradictory signals over a period of several days. This involved changing a vessel name twice, as well as falsifying its identity code and intended destination.

Elsewhere, criminals are using satellite technology to help navigation and location, not just to avoid detection. Narcotics smugglers have for some time been known to utilize GPS technology to aid their activities. Quantities of drugs are left at sea, often on floating devices weighted so that they stay just below the surface to avoid optical detection. However, the packages are also attached to buoys with embedded satellite location devices, which transmit their location on specific frequencies to allow the recipients of the goods to find and collect them. This method increases the flexibility available to the criminals, in case either of the vessels are being tracked – they never actually have to meet to transfer the drugs.

If confirmation were needed of the willingness and ability of drug traffickers to utilize maritime technology to the full, it probably comes with the understanding that such criminal networks have been deploying their own fully functioning submarines (soon labelled as 'narco-submarines') in the past few years, which provide them with the capacity to transfer far larger shipments than on disguised fishing vessels. Measuring up to 15 metres, they are made of fibreglass and synthetic materials, and specifically designed to minimize the chance of detection by radar or infrared. Estimated to cost approximately £750,000 to build, these diesel-electric subs – some of which are fully air-conditioned – are abandoned after they have reached their intended destination.

Such investment in ocean-going technology by cash-rich criminal networks only goes to underline the need for security agencies to continue investing in ever-more innovative counter measures that are not solely restricted to investment in satellite-enabled Earth observation systems.

Robocams

As with most Blue Economy sectors, satellite systems are also complemented by the emergence of autonomous and unmanned systems on the Earth's surface in the maritime surveillance domain. While offering great leaps forward in maritime surveillance, data from satellite-based Earth observation systems can still usefully be merged with information collected at sea level.

Enforcement agencies have long used photographic imagery in their work, but the development of miniaturized and ruggedized still and video cameras that can be powered on small unmanned systems at sea is progressing at a rate of knots. This means that high-quality, usable images can be taken and relayed to shore-based authorities, without the need to launch expensive manned missions. Additionally, the image capture will have a date and time stamped on it, as well as precise location data, all very useful as evidence in any resulting court cases, or for relative coastal agencies to have enough confidence to task additional assets in the area.

Small autonomous or remotely operated systems have another key advantage over larger, manned surveillance patrol vessels – stealth. An unmanned surface vessel (USV) as small as 5 metres long is incredibly hard to pick out in a vast expanse of water, so seafarers undertaking nefarious activity would not feel the need to hide their actions (be that hiding enslaved staff, transhipping contraband or using gear forbidden in certain fishing zones) in the same way as they might if they saw a larger vessel. In this respect, unmanned systems that are silent (powered by waves or solar energy) provide the greatest assurance that their important covert element will not be compromised.

Sea-level autonomous and remotely operated vessels can provide much more data than the imagery and positioning information provided by satellites. Passive acoustic monitoring sensors could be used to assess noise pollution on construction projects in sensitive areas; heat sensors could identify human activity on board; other sea-level sensors could detect changes in the water (eg potential oil pollution on dredging projects or activities further out to sea where operators believe they are 'out of sight, out of mind').

And, as autonomous systems such as those that utilize wave or solar power are not troubled by the need to return to shore to refuel or to change crews (though they do need occasional de-fouling), they allow wide-range and persistent data gathering. Once fleets of such vessels are being employed within specific marine spaces, the net will be tighter around those wishing to commit illegal activities.

Quality, quantity and safety of data

Some enforcement agencies are beginning to warm to the potential of autonomous and remotely operated vessels as a new tool in the maritime domain. Of course, data transmission permitting, they also allow real-time transfer of information, which can be acted on immediately. This leads to one critical issue related to autonomous maritime surveillance (as, indeed, with surveillance from space) – that the quality of the data is sufficient to either drive more costly interventions, or for use in civil or criminal prosecutions at a later date. Enforcement monitoring at sea is intelligence-led, with information emanating from a variety of sources. No coastguard or related authority will want to task a naval vessel to travel to a specific area to investigate a potential issue based on intelligence about which they are uncertain. So, they may ask themselves, can you rely on video evidence from an unmanned vessel if it's been operating in a rough sea state and poor visibility? As with so many emerging technologies in the Blue Economy, such questions and concerns will only be answered with the further testing and deployment of such systems, and the wider sharing of any resulting case studies within relevant stakeholder communities.

Another issue involving autonomous systems being utilized for maritime surveillance activities relates to the safety and security of the systems themselves. As mentioned elsewhere, one of the greatest attractions of autonomous sea-going vessels is that they remove the need to send humans out to sea, which is especially important when considering deployments in harsh ocean environments. However, that does not remove all risk to the technology itself. Natural and environmental risks affecting autonomous vessels include losing stability due to biofouling (eg barnacles and algae clogging up essential systems), collisions, getting caught in fishing nets, losing connectivity, and more. Within the maritime surveillance realm, however, these threats are exacerbated by wilful human intervention from criminals desperate to avoid detection. Autonomous and remotely operated systems can try to minimize risk here by choosing camouflage paint options, but this can only work to a certain degree.

One unfortunate example of this comes in the illegal fishing domain. In January 2018, environmental conservation group Sea Shepherd reported that an aerial drone that it was operating had been shot down by poachers. The drone was being deployed by the NGO within the Gulf of California, and its specific mission was to monitor illegal activities that pose a threat to the critically endangered vaquita porpoise. The flying drone was fitted with a video camera that was transmitting in real-time to its operators, which meant that the footage could subsequently be distributed far and wide. It captured a fisherman in a small speedboat – clearly not happy to be followed – firing repeatedly at the drone, which eventually succumbed to the attack.

At around the same time, the navy of Yemen's Houthi Ansarullah movement released a jubilant video online showing what they claimed to be the recent capture of a US spy drone. A US defence spokesperson soon after identified the device as an unmanned underwater vehicle carrying out meteorological research, but the fact that it had been captured and removed from its operating environment underlined that – whether deployed at sea or in the air – unmanned or remotely operated vessels are by their very nature vulnerable to aggression.

However, the flipside of this potential downside also reconfirms one of these devices' greatest strengths, in that they completely remove the risk of harm to human operatives. Whether deployed in high-risk or environmentally harsh areas, no matter what happens to them, the risk to human operatives is zero. While no human operatives may be at risk, however, any data collected by autonomous or remotely operated vessels remains vulnerable to physical capture. As with every ocean-going data operation, as discussed in greater detail in Chapter 2, this is compounded by the possibility of cyber interference. While perhaps not as obvious a target as large commercial or military vessels, the cyber protection available to sea-going autonomous vessels is obviously restricted to the size and complexity of the equipment it can carry, which itself is severely limited by available power. Perhaps more of a concern for state actors than commercial enterprises, developers of such systems have their work cut out to devise systems that minimize this risk, especially when one option – to transfer all data immediately and automatically to the cloud via satellite communication – remains very expensive.

Some enforcement professionals also remain unconvinced at the speed at which unmanned systems can patrol wide areas. While the ability to travel at double-digit knots is desirable, it is not currently offered within the marketplace on vessels that can travel persistently over large expanses of water for many days or weeks. This, of course, is where complementarity between systems in the maritime surveillance domain becomes important. If satellite tracking and imagery can provide data on areas of greatest potential in relation to suspicious activity, or if a known vessel of interest is seen heading in a specific direction, unmanned vessels can be sent to such 'honeypot areas', or to intercept a known target within range, to complement such intelligence by providing sea-level data.

Finally, the continued progression of unmanned systems within the maritime surveillance domain presents another challenge if true scalability is to be achieved – that of how to manage huge volumes of data in a way that boosts rather than swamps management systems. The complexity of such data is one starting point to consider. As well as providing evidence that a vessel is identifiable, cameras within the fisheries domain can capture types of gear being deployed into the water (and that might just mean wires leading off the vessel in to the sea, which will indicate illegal activity within that particular seaspace); imaging can also capture the faces of individuals; sidescan sonar systems can track what is happening under the surface of the waves; the list could go on. Each of these creates new opportunities and – as ever, as cutting-edge technology progresses at pace – new challenges to be met to enable the fullest return.

For example, one of the key benefits of autonomous systems collecting data at sea is the fact that you do not need to operate them 24/7, so staffing resource costs are drastically reduced. However, if such systems are tasked to deliver maritime surveillance, and provide near-constant data streams via satellite, is that of any use if not viewed immediately by an analyst who can view and respond accordingly within a timeframe that allows potentially criminal activity to be investigated while it's taking place? So yet again in the ever-expanding world of Big Data and artificial intelligence, we come back to the need to develop complex algorithms to ingest and understand new information and, for example, only create an alert for human review or intervention should a specific incident be noted. Visual recognition software could, for example, identify a vessel of a certain type from a video feed to trigger a shore-based analyst to review the footage. Such automated systems, it should be noted, are being enthusiastically developed in many Blue Economy areas, as is evident in the EASOS Marine Watch example above, and in many other chapters of this book.

Conclusion

As well as providing new and compelling evidence of the availability, trustworthiness, affordability and complementary nature of maritime surveillance data newly available from space (both via the new wave of smallsats or more traditional launches) and sensor-laden unmanned systems on the Earth's surface, Blue Economy entrepreneurs also need to collaborate to develop systems to cope with all this new data in ways that lead to actionable insights.

Maritime surveillance is driven by a broad range of challenges, from tracking climate change to identifying and countering the illegal passage of all forms of contraband. Those challenges are, of course, as large and complicated as the oceans are wide, and made more complicated still when the vast majority of the noticeable business at sea is safe, legal and part of the quotidian ocean-going activity that underpins so much of the world's trade, transport and leisure.

Against this backdrop, identifying crime at sea can be like looking for a needle in the world's largest, continually moving haystack.

Satellite positioning, navigation, Earth observation and communications are key enabling technologies within the maritime surveillance domain, with unmanned systems also offering the promise of new capabilities and much more capacity at reduced cost. Whereas until relatively recently satellite data may have only been available intermittently, the proliferation of large and smaller satellites provides those monitoring activities on the seas and oceans with daily updates to pipe into their ever-more sophisticated systems. However, the rich and voluminous data provided can run the risk of making grappling with surveillance more difficult, rather than driving more intelligence-led, targeted actions. If identifying crime at sea is indeed like looking for a needle in a haystack, the last thing a regulator or enforcer needs is another 100 haystacks to consider.

Development of machine learning algorithms to crunch data may not originally have been uppermost in the mind of innovators who designed unmanned surface vessels for maritime surveillance purposes, but it is one of a number of issues that need to be progressed in tandem to maximize the market potential of their own inventions. While the response to that challenge may be considered internally, with the development of new bespoke data services, partnership with other companies with data and AI specialisms offers perhaps a more sensible route to meeting the user's need. Such cross-cutting collaboration is mutually beneficial, of course, as Big Data and machine learning innovators are always looking for new needs in untapped domains.

We come back to how such developments are to be funded and how long that may take to happen, which is explored in greater detail in other chapters on funding. What is clear, though, is that the sensible innovator will consider manageable data analysis as both an opportunity and a challenge to be countered within a comprehensive business plan that considers major barriers to market success.

As sea-borne criminals attempt to channel technological advances to their own ends, maritime surveillance challenges can increase in difficulty. The tech-enabled maritime surveillance prize grows ever larger, then, but so do its development needs. Thankfully, when put in the context of the size of crime at sea (itself only one element of maritime surveillance), a number of independent experts have been expressing in forceful terms that the high revenue potential of the use of technology to, for example, counter illegal fuel smuggling will significantly outweigh the low cost of the initial technology investment. This problem is therefore an 'invest to save' model. Continuing with the fuel smuggling example, the market desire for discounts on fuel is the fundamental driver of this illicit trade; this is not a driver that will recede without strong intervention. The future therefore relies on continued collaboration between innovators in the burgeoning smallsat domain, maritime engineering, sensor development initiatives (especially those that progress miniaturization of sensors that can be carried on unmanned systems with relatively limited power options) and data innovators who can marshal computing power to provide new forms of actionable intelligence to those attempting to maintain vigilance over our increasingly complex seas and oceans.

References

NSR (2018) Small satellites flying high with $37 billion market and 6,500 satellites to launch by 2027. www.nsr.com/small-satellites-flying-high-with-37-billion-market-and-6500-satellites-to-launch-by-2027/ (archived at https://perma.cc/4LX6-JY76)

Russell, K (2017) Situational awareness: An opportunity for smallsats. www.satellitetoday.com/innovation/2017/03/28/situational-awareness-opportunity-smallsats/ (archived at https://perma.cc/7KQM-5BJE)

Wilson, B (2015) Five maritime security developments that will resonate for a generation. https://harvardnsj.org/2015/03/five-maritime-security-developments-that-will-resonate-for-a-generation/ (archived at https://perma.cc/BTR9-WFAQ)

07

Aquaculture

Meeting the world's food needs in an environmentally sound manner

The tipping point came in 2009. That was the year when it was revealed that the majority of fish finding its way onto the world's plates came from controlled farm environments, rather than being caught in the wild. Aquaculture production nearly tripled in volume between 1995 and 2007, and has continued at pace ever since. The authors of the report highlighted the public's demand for the omega-3 fatty acids found in oily fish (believed to be good for the heart) as one of the key drivers behind this meteoric rise of farmed fish (Livescience. com, 2009), but this also aligns to the growing basic need for fish as a food source, the background to which we explore in greater detail in Chapter 10, which focuses on sustainable fisheries. However, aquaculture deserves greater investigation, due to its growth prospects and the particular challenges and opportunities that it presents.

By 2016, production from aquaculture had reached 80 million tonnes, providing 53 per cent of all fish consumed by humans as food (Food and Agriculture Organization (FAO), 2018) and the World Bank estimates that the size and scale of the aquaculture market will continue to grow until by 2030 it will account for 62 per cent of all the seafood we consume (World Bank, 2014). This is not only important news for the Blue Economy; the FAO has also pointed out that aquaculture continues to outpace every other food production sector, growing as it has at an annual compound growth rate of nearly 6 per cent since

2010 (US Soybean Export Council, 2018); in 2016 alone, the aquaculture sector increases production by some 4 million tonnes year on year. Additionally, in its comprehensive 2016 Ocean economy report, the Organisation for Economic Co-operation and Development (OECD) highlighted the industrial marine aquaculture market as one of the key prospects for high long-term growth, with employment within the sector predicted to increase by over 150 per cent by 2030.

Not surprisingly, the economic statistics that accompany such predictions also look healthy. The aquaculture market – valued in 2018 at £143 billion – is anticipated to reach £178 billion by 2022. This will be achieved by successive compound annual growth rates of 4.12 per cent (2019); 4.50 per cent (2020); 4.83 per cent (2021); and 5.15 per cent (2022) (White, 2018).

Many countries are already heavily reliant on aquaculture. The industry accounts for over 50 per cent of Philippine fisheries output, for example (in contrast to European Union markets, where aquaculture accounts for approximately 25 per cent of overall output), and is regularly the one market sector whose steady growth allows the country's overall fisheries output statistics to retain some buoyancy, as standard commercial and municipal fisheries can fluctuate significantly (Psa.gov.ph, 2019). Over 50 per cent of the country's animal protein consumption comes from fish, so aquaculture is hugely important in providing the most basic of nutrition to the citizens of a nation that still registers significant involuntary hunger problems from time to time (Cabico, 2018).

There is no doubt, then, that substantial growth in aquaculture appears to be a safe bet. An increasing population means growing demand for protein, and so the market is expanding, and with it the opportunity for financial gain. With such compelling statistics, it is no surprise that countries and businesses around the world are taking great interest in how they can support and benefit from an industry that is universally accepted to offer high growth in the coming decades.

In Southeast Asia, Vietnam's authorities are facing up to the opportunity as they recently revealed the country's ambitious intentions to become the world's top aquaculture producer (Dao, 2018), with a headline aim of achieving aquaculture export value of £7.7 billion by the year 2050. Achieving top spot in the global aquaculture rankings will mean overtaking the current global top three of China, Indonesia and India. It is obvious that such ambition cannot possibly be reached without significant investment, and fisheries leaders are certainly doing so. Taking one example, the Vietnam Association of Seafood Exporters and Producers announced in 2018 that the central coastal province of Phu Yen will invest nearly 2.12 trillion Vietnamese dong (approximately £72 million) to help develop its aquaculture industry up to 2025.

Like many Southeast Asian countries, the majority of its recorded aquaculture output actually comes from small-scale independent fish farmers – in Vietnam's case, involving some 50,000 households. While these producers tend to use traditional methods and very basic equipment, the industrial aquaculture race is also driving demand for new and innovative technology-led practices.

In October 2017, the UK's Centre for Environment, Fisheries and Aquaculture Science and the University of Exeter united to launch a new aquaculture research centre to catalyse progress in sea animal health, food safety and protection of the aquatic environment (Undercurrent News, 2017). The Collaborative Centre for Sustainable Aquaculture Futures was launched by the UK Government's Environment Secretary Michael Gove and its initial areas of investigation were listed as including aquatic disease; anti-microbial resistance; aquatic disease modelling and epidemiology; and environment and animal health.

Scotland has long been a country in harmony with the seas and oceans. From oil and gas to fisheries, coastal tourism and more recently managed aquaculture, Scotland's Blue Economy is entwined with its unique culture, topography and ocean space. So, it is no surprise that Scotland is offering significant leadership to explore the aquaculture opportunity. In 2015, 96 per cent of the UK aquaculture sector and all the UK salmon farming industry were in the Highlands and Islands of Scotland. This industry provides a significant contribution to the economic and social fabric of the region, especially for people who live in very remote communities. The aquaculture industry contributes over £1.8 billion to the Scottish economy and accounts for a staggering 7 per cent of global production. This market share has declined from 10 per cent in 2005; global production is increasing at such a rate that the Scottish market share is not keeping pace with the rest of the world. Unless technology and improved processes can be exploited to increase production, Scotland could fall further behind as global production continues to grow at approximately 5 per cent year on year. To galvanize action, Scotland has, like Vietnam, set itself ambitious growth targets, aiming to double its value from £1.8 billion currently to £3.6 billion by 2030 (Gatward et al, 2019). While the industry is not looking to increase overall tonnage, it needs to maintain the value through a high-quality product, with sound provenance.

To underline the potential of fast returns here, the Iceland statistics authority was able to release an astonishing figure in January 2019, as the country's aquaculture sector was announced to have increased its operating income by 34 per cent year on year, with associated pre-tax profits increasing by 2.6 billion Icelandic kroner (£16.6 million) (Hjul, 2019). This continued a growth trend that meant that the industry's earnings had more than doubled in the previous three years.

The global race

None of these eye-grabbing targets will be achievable with traditional methods, though, so the global race for technology innovation – in the West and the East – is very much on, and innovative applications across all elements of aquaculture development and delivery are beginning to attract investment. Achieving ambitious growth targets within the aquaculture industry will require more accurate monitoring to maintain stock health, and therefore value, and a better understanding of the environmental impacts to ensure the industry can grow sustainably.

Before that, though, the very first thing you need to do with a new fish farm is know where to locate it. As with many Blue Economy sectors, satellite-enabled systems are beginning to offer value here.

As an example, TCarta (featured in greater detail in Chapter 8 on hydrography and bathymetry) delivered a ground-breaking project in the Arabian Gulf in 2018, as they put their satellite derived bathymetry data to good use (Pobonline.com, 2018). BMT, an innovative UK engineering and scientific consultancy firm, utilized specialist TCarta data sets to aid in the selection of new fish farming sites in the area, a service they are providing to the Abu Dhabi Environment Agency. Companies collecting satellite-derived bathymetry have been growing their databases in the past few years, both to help to address the major dearth of accurate seafloor depth measurements globally, and to open up new commercial and environmental opportunities that such high-resolution satellite imagery makes possible. In this instance, hydrographic modelling software helped to pinpoint ideal fish farming sites based on water depth thresholds (to accommodate fish cages) and alignment with natural subsurface channels whose water currents can flush away waste.

As both the TCarta/BMT and the varied focus of the new aquaculture research centre suggest, potential environmental impacts are at the forefront of aquaculture development, and despite the size and relative maturity of the industry, these disciplines are in the early stages of being truly understood. There are three sides to this particular coin: having the right natural environmental conditions in place for fish to thrive; being able to monitor those conditions to ensure they remain stable as – by deteriorating – they would threaten the health of the cultivated fish stock; and finally, making sure that the aquaculture site does not harm the surrounding environment.

Taking the middle of those issues, aquaculture managers live in constant fear of an outbreak of disease that can threaten their stocks. Healthy fish require monitoring. Adherence to stringent environmental and health standards is essential for responsible food producers. Some of the major factors to be considered in aquaculture are animal health, hygiene and particularly water use. Disease and sickness can spread easily among the animals if not identified swiftly, and because of the nature of these floating aquatic farms, the outbreaks can also spread to wild populations, putting sites at risk of immediate closure.

Fish farmers in Lake Toba, the biggest lake in Indonesia, have suffered more than most in recent years in relation to these issues, with two separate incidents (one in 2016 and one in 2018) devastating their operations and leaving millions of fish dead on each occasion (Karokaro, 2018). Perhaps the most worrying aspect here is that both incidents came with almost no warning. While some fish farmers in one of the incidents reported that they had noticed some unusual activity within their floating nets, for many it was a bolt from the blue, sadly giving them no time at all to respond.

These examples have been attributed to a sudden depletion of oxygen in the water. Such conditions can come about due to a toxic mix of pollutants building up in the lake, unfavourable weather conditions and – ironically – poor practice from local fish farmers whose operations reduce the water quality. Not being able to sustain such environmental and economic shocks on a regular basis, the Indonesian authorities are taking measures to help counter the possibility of such incidents happening again. In October 2018, fisheries officials launched a predictive calendar to alert fish farmers in Lake Toba to the emergence of water conditions that may prove threatening to their fish stocks (Gokkon, 2018). The alerts are intended to ensure that fish farmers react positively to worrying signs of water health by, for example, reducing feed amounts that can in turn exacerbate emerging problems.

However, such planning tools (sensible as they are and even if available online) are just one of the more basic options available; others utilize the power of machine learning to make a difference. For example, the aquaculture health response system being developed by analytics company Manolin aggregates large amounts of data and quickly notifies fish farmers when an outbreak of sea-lice is emerging in their region (Fletcher, 2018). As with any Big Data project, there are a number of problems that a unified system can solve: there's too much data available to make proper sense of, it can be challenging to aggregate it from multiple sources on multiple systems, a lot of data is actually on paper, etc.

The Manolin team – based in Bergen, Norway – are developing an automated system to get key information to interested operators much more efficiently. They may have a willing market when you consider that it has been estimated that sea lice now impact 17 per cent of all aquaculture production costs, an impact that has grown 233 per cent since 2010. This is very much a problem in need of a rapid solution, and especially in Norway, where the annual loss of stock caused by sea lice is estimated to be £500 million. It is worth considering the current manner of operations to better understand the potentially transformative effect of services such as those offered by Manolin. Salmon farmers are obliged to undertake weekly surveys of their cages to measure sea lice levels and report findings to a central authority that collects the data. In many cases, the current method is to manually check 10 fish in a cage of about 200,000 specimens. Of course, this rate of sampling is not statistically significant, so serious vulnerabilities remain.

In full bloom

Another issue that fish farmers have increasingly had to face up to in recent years is the further rise of harmful algal blooms (HABs). These environmental incidents occur when colonies of algae – simple plants that live in the sea and freshwater – grow out of control and then produce toxic or harmful effects on people, fish, shellfish, marine mammals and birds. While we know of many factors that may contribute to HABs, how all the relevant factors come together to create a 'bloom' of algae is not well understood, though studies indicate that many algal species flourishes when wind and water currents are favourable. In other cases, HABs may occur when nutrients (mainly phosphorus, nitrogen and carbon) from sources such as lawns and farmlands flow downriver to the sea and build up at a rate that 'over-feeds' the algae that exist normally in the environment. Some HABs have also been reported in the aftermath of natural phenomena like sluggish water circulation, unusually high water temperatures, and extreme weather events such as hurricanes, floods and droughts.

The economic impact of HABs in the United States at the beginning of the millennium was estimated at £77 million per annum (Anderson et al, 2000). Humans can get sick by eating shellfish containing toxins produced by these algae. Airborne HAB toxins may also cause breathing problems and, in some cases, trigger asthma attacks in susceptible individuals. And, of course, they represent a severe threat to the captive colonies of fish contained in aquaculture sites. In March 2018, over 4,000 tonnes of stock were lost at Chilean salmon farms due to a HAB event (Garcés, 2018). Two years previously some 25 million fish (20 per cent of total farmed stock) were killed by HABs off the Chilean coast, costing the industry £565m (*Guardian*, 2016).

To counter environmental threats such as those offered up by HABs, manual sampling of water, manual analysis and manual delivery of outputs also takes place at aquaculture sites. Fin fish farms will monitor water quality prior to feeding and treating fish to ensure minimum chance of harm to stock: this is time-consuming and is often not timely enough to prevent stock damage. Shell fish farms monitor water quality prior to harvesting, although subsequent quality checks on stock at the processing plant can often highlight the presence of biotoxins previously not identified.

One of the most exciting developments in the aquaculture sector is the combination of new, powerful imaging sensors with the emerging capacity to crunch data quickly via machine learning to provide new forms of value. Aquabyte is one such example. Based in Norway and the USA, it is an emerging company developing machine learning and data visualization capabilities for fish farmers, promising to identify potential problems and suggest ideal feed flows – ultimately, harnessing data to increase yields. By installing underwater 3D cameras in the farm's net pen, the company promises four distinct areas of added value: lice counting (to enable pre-emptive decision making in case of an outbreak); biomass estimation (to allow accurate projections); appetite detection (to improve rates of feed conversion at the same time as reducing waste); and feed optimization (using recorded fish size and activity to suggest recommended feed amounts). The company suggests that by applying its machine learning algorithms to the imagery collected underwater, more efficient feeding over a fish's lifetime could result in as much as a 30 per cent decrease in feed cost. Across a market as sizeable as aquaculture, that could lead to tens of billions of dollars saved in the coming years.

Communication breakdown

Many of the inspection and monitoring approaches being developed are retrospective rather than predictive or preventative, and even those that are forward-looking are also bumping up against one of the broader cross-cutting issues affecting the sector – reliable and affordable connectivity. As the Lake Toba examples show, by the time a problem becomes noticeable by standard inspection methods, it may already be too late. A lack of digital connectivity also means that acquiring and collating water quality measurements from multiple sites is incredibly time-consuming, further adding to the monitoring burden of the industry. This is, we believe, one of the major issues currently facing the aquaculture industry: a lack of digital connectivity means that many sites are still working in an analogue world. There is little incentive for automated data collection when it must be manually sent, by road, and it is often too late to action when it arrives at its end location.

Jumping back across the North Sea to Scotland, the home of UK aquaculture, services do exist to provide one-off environmental impact assessments and in-house monitoring takes place on individual sites. The aquaculture industry in Shetland, for example, currently makes use of an online water quality bulletin service called HAB Report, which is produced weekly by the excellent experts at the very forward-thinking Scottish Association of Marine Science. This service makes use of free-at-point-of-use satellite data to provide a forecast of biotoxins in the waters in and around Shetland. It is paid for by the National Environmental Research Council and the Biotechnology and Biological Sciences Research Council, with the project team manually producing the report. The project funds do not cover the costs of purchasing additional satellite data, to increase spatial resolution of forecasts, nor can the team produce the reports more regularly, meaning its temporal resolution is understandably limited. Again, the lack of joined-up systems means there is no feedback loop for the HAB Report service, so the bulletin providers are not able to confirm how accurate their predictions were, nor do they have access to in-situ data to enhance the spatial resolution of forecasting.

The repeated barriers to progress presented by a lack of connectivity at fish farms is one of the reasons why the authors are collaborating with an array of stakeholders in Scotland to explore the potential of a smart, connected fin/shellfish farm. Supported by the European Space Agency, NLAI Ltd is working with the Scottish Centre of Excellence in Satellite Applications, located within the University of Strathclyde, to develop a system that allows aquaculture operators to understand in greater granularity what impact the environment is having on their stock, and what impact they are having on the environment.

The emerging system is anticipated to utilize satellite Earth observation combined with in-situ sensors to assess the water conditions that matter most (eg salinity, turbidity, temperature) and provide ground truth for early warning mechanisms for biological threats such as HABs. A low-power wide area network for machine-to-machine communication and data transfer to the shoreline will be developed, to counter the data transfer issue. This Internet of Things system will also collect data on fish health and behaviour, feed and medicine usage, etc. Finally, a command and control system will be developed with appropriate analytics and machine learning to enable automation of algal species detection, coupled with hydrodynamic modelling and prediction for better decision making.

We have decided to initiate this project as aquaculture is so clearly going to be an important cornerstone of future Blue Economy activity, and connected services within the sector are still in their infancy. Within the UK, when poor water quality is detected this can result in the short-term closure of farms, costing a shellfish business up to £162,000 per closure due to loss of stock and sales (Shelleye.org, 2017), so any new system that can accelerate the dissemination of valuable data has obvious worth.

Others are also working hard to progress these cross-cutting issues. In New Zealand, the Precision Farming for Aquaculture project is focusing on radically reducing maintenance costs by introducing chemical and imaging sensors to allow fish farmers to monitor conditions remotely from a computer or mobile device (Fletcher, 2017). The project aims to utilize sensing technologies, lasers and artificial intelligence to reduce the need for farms to be physically accessed by boat to manually record stock health and condition. As well as acknowledging the complex connectivity issue, it is also exploring efficient and cost-effective underwater communications.

Net effect

As well as monitoring the general health and environment of the water and the fish being cultivated, there is also a need to protect most important piece of equipment in the set-up – the aquaculture cage/net itself. Marine growth build-up on aquaculture nets is one of the main identified problems in fish farming. Unchecked, this can lead to a lack of oxygen and the subsequent growth of parasites, bacteria and shellfish. More immediately noticeable is the fact that nets that become heavy due to the amount of biofouling they bear can lead to the risk of the net breaking and stock escaping or predators entering the farm – both of which can have very serious consequences.

One of the most devastating examples of this in recent years happened at a farm off Huar Island in the Patagonia region of southern Chile in July 2018. Salmon cages were breached after they had been battered by a raging storm. This led to an astonishing 690,000 salmon being released into the open ocean (Salazar, 2018). As well as obviously being economically devastating for the operator, there are also potentially greater environmental risks as the escaped farm salmon – a natural predator – interact with the native species. They may attack smaller species, affecting the local ecosystem (especially when they enter it in large numbers), or they may breed with wild salmon, interfering with their genetic make-up.

A similar incident in Tasmania in May 2018 not only had similarly troublesome economic and environmental effects, but also led to what might well be lasting damage for the image of the aquaculture sector in the region. Again caused by a violent storm – Hobart's worst for decades that saw over 100 millimetres of rain in 24 hours seriously disrupt infrastructure and cause millions of dollars of damage

(Glumac, 2018) – an industrial fish-feeding machine inside a pen owned and operated by Huon Aquaculture smashed through the pen (which eventually washed up on a beach), thus creating a gaping escape route for the pen's fish stock. Like the case in Patagonia, this incident threatened the local ecosystem as the previously contained salmon learned to hunt and started to take on the native species, so locals were obviously keen to understand the full extent of the case. However, the company initially appeared to resist announcing how many fish had escaped (despite, it was claimed, providing such information to government authorities), but were eventually apparently 'bounced' into admitting that up to 120,000 fish had escaped the pen by a journalist from ABC News (Compton, 2018). The aftermath of the event, then, at least as it played out in the media, appeared to focus more on the need for transparency rather than the actual damage caused.

While, as we have seen, there is a huge role for aquaculture to play in providing the world's protein needs in the decades to come, many environmentalists are concerned about the industrialization of fish farming and the effects it may have on water quality, as well as the impacts of escape incidents. As it has also been predicted that climate change is likely to increase the number of extreme weather events at sea in the coming decades, achieving greater pen stability is an issue that undoubtedly needs to be grasped. The ideal is not to lose any fish at all, and innovators have been applying themselves to solving the problem.

The growth of biofouling on nets depends on location and weather conditions, and is traditionally countered by regular mechanical cleaning, and/or the use of specialist anti-fouling paints or chemicals. New anti-fouling materials for nets are also being developed. Mechanical cleaning involves the use of a high-pressure cleaning system, which requires a large operating platform (often a ship) and substantial costs in man hours and time, so innovations that reduce costs are being sought.

Multi Pump Innovation is a Norwegian technology company that has brought an innovative net cleaning robot successfully to market (Undercurrent News, 2018). They claim that their Racemaster 3.0 system is the world's fastest net cleaning robot, and have sold the system to customers in Norway, Scotland, Australia, Canada and Chile. Controlled manually by an operator above water and propelled by nine water jets, the system cleans 1.8 metres of net at a time with three rotating discs and has the flexibility to cope with loose and tight netting. The system operators follow progress via hull-mounted cameras, so can see, for example, if/when the system needs to be righted if it has flipped over. Being able to perform such corrective action immediately thanks to real-time visuals obviously enhances operational efficiency, and in turn reduces the potential for netting becoming overly fouled and therefore at greater risk of breaking.

Imprisoned in steel

Another bold claim has been made by the Norwegian developers of the Aquatraz, who claim that they are developing an 'escape-proof' fish cage (Phys.org, 2019). Its rigid steel construction emboldened the manufacturers enough to take the inspiration for its name from San Francisco Bay's notorious high-security prison Alcatraz. The first cage in production is 18 metres deep with a circumference of 160 metres, with the company also promising additional health and welfare benefits to fish resulting from an innovative water pump system.

Many of the key risks to aquaculture development featured here – including susceptibility to sea lice and being exposed to severe weather events – arise due to the need to locate fish farm pens close to the surface of the water. These risks, then, become neutralized if aquaculture pens do not have to operate at surface level. That is exactly the premise behind the solution being developed by Mowi, another innovative Norwegian company. Their AquaStorm project will initially see pens positioned 15 metres below the surface of the sea – out of the range of sea lice and protected from the more severe fluctuations experienced at surface level during heavy storms. While still applying for several licences to push ahead, the company suggests that in time they envisage being able to sink AquaStorm nets to a depth of 50 metres for even greater stability. Similar, the first installations are intended to be positioned up to 12 kilometres from the coast, though they believe in theory that such operations could eventually be installed up to 100 kilometres from land. An onshore control centre will allow remote management of feeding cycles and monitoring of environmental conditions.

Mowi is the world's largest provider of farm-raised salmon, but it is clear that they understand the need to innovate to meet the kind of growth targets referenced throughout this chapter, as do many of the larger companies, especially within such exciting collaborative endeavours as the Bergen Seafood Innovation cluster. This one ambitious project alone has brought together 57 partners to focus in on key development challenges and bring new innovations to market.

Similar critical themes as they relate to innovation are emerging across the sector. For example, in November 2018 the US Government's National Oceanic and Atmospheric Administration (NOAA) published the latest investments they had made into aquaculture research (Carney, 2018): $11 million in total was distributed through the very popular Sea Grant programme, to support 22 individual three-year projects that focused on priorities such as: the development of new technologies; communicating accurate, science-based information about the benefits and risks of marine aquaculture to the public; and increasing the resiliency of aquaculture systems to natural hazards and changing environmental conditions. The projects include innovation in sea lice management, exploring how to improve the feed conversion ratio, and developing systems to assess water quality from space-based assets.

Conclusion

Our research in aquaculture sees a development and growth in both plant size and design. Everyone is gearing up for expansion, the race for growth is undoubtedly on, but the aquaculture industry needs to think long-term. In fact, the sector could be seen to be at a crossroads. While market growth statistics are compelling across the piece, many market analysts predict that such growth will be achieved partly by focusing on polyculture (the simultaneous cultivation of several species at the same time) and through the deployment of greater intensification methods. If not managed well, these approaches could do more harm than good in the longer term. The industry's capitalization needs to live alongside the needs of societies, cultures and ecology to ensure that all three are sustained in the long term. As evidenced throughout this book, Blue Economy solutions – when implemented well – ensure environmental considerations are met while engaging local societies in the process and ensuring that the prosperity that is created is fairly distributed.

A cross-cutting 2017 report on Norwegian aquaculture from multinational professional services firm PricewaterhouseCoopers highlighted many issues facing the sector. For example, global warming could have a major impact on the potential of aquaculture development in the coming decades (PwC, 2017). There are numerous other potential threats; for example, warmer sea temperatures may create conditions more conducive to bacteria and viral diseases, and the rising prevalence of plastics and microplastics in the sea will provide severe challenges to the marine food chain. More generally, though, the PwC report was notable for how often it referenced the issue of protecting the industry's reputation through the promotion of healthy and sustainable practices as being of paramount importance.

So, the many countries setting ambitious targets to double fish farming output or to become the top aquaculture provider in their region or across the world must also embrace the challenge of sustaining that status in the long term through research and innovative implementation of technologies as well as developing a thorough understanding of social and environmental impacts.

There are more than murmurings about the potential negative effects of aquaculture. In November 2018, the Norwegian city of Tromsø announced that it was putting severe restrictions on the fish farming industry within its municipal boundaries, to the horror of the local aquaculture sector (Hjul, 2018). New sites were to be banned, and existing at-sea facilities were not allowed to seek expansion. While the council eventually backed away from the severity of its decision to ban new activities outright some five months later (Hjul, 2019b), it is a shot across the bows that the aquaculture industry will no doubt have noticed across the globe.

Additionally, striking amongst the award winners of the 2018 Sea Grant National Aquaculture Initiative from NOAA were two projects that underlined the importance of this issue. One focused on improving the public perception of farmed finfish (which was deemed to currently be largely negative); another focused on increasing end-user confidence in US farm-raised seafood in general.

It's important to point out that the debate is not entirely overshadowed by reflection of negatives. The more positive aspects were highlighted, for example, in a research paper in the journal *BioScience* in December 2018 (Thefishsite.com, 2018). Within the report, a team of marine scientists from the University of Adelaide, The Nature Conservancy and Macquarie University took the opportunity to underline the broader benefits – aside even from the provision of much-needed animal protein capacity – that a growing aquaculture sector is well positioned to deliver. Medicines and services such as water treatment, shelter/habitat for wildlife and prevention of erosion were all explored. However, the very fact that the headlines associated with the report led with the assertion that the benefits of aquaculture may be being overlooked illustrates how far the pendulum of opinion may have swung. As with the PwC report, the authors urged the industry to be more conscious of the design of their installations and their interactions with their surroundings, to enable them to maximize the positive effects.

Of course, technology has a vital role to play in achieving the appropriate equilibrium within the sector. Indeed, it is notable that the council in Tromsø only agreed to water down its original banning proposals as long as the local aquaculture industry agreed to embrace technology that promoted outcomes that were more environmentally, climate change and socially friendly.

NLAI's own aquaculture research continues with concepts to trial environmental monitoring using technologies such as space-based Earth observation, machine learning and the underwater Internet of Things. Our regular conversations with members of the aquaculture industry, especially in northern Europe, point to a desire for shared, open data from *in-situ* and remote sensing to enable continuous monitoring of the marine environment across the entire region. The development of such systems will not only provide faster dissemination of critical environmental information but will also open up the potential for a more robust feedback loop, thus enhancing the information services provided. Of course, the more data that becomes available, the greater the potential to spot trends and identify opportunities for more impactful interventions and enhancements with multiple benefits.

It should come as no surprise that China harbours significant aspirations here. The country dominates the fisheries sector, and in January 2019 announced the launch of the China Intelligent Fisheries Association, intended to bring data specialists, fishing companies and government officials together to work out the best ways to harness Big Data and artificial intelligence capabilities in order to drive innovation within both the country's aquaculture and wild-catch fisheries (Godfrey, 2019). Focusing specifically on the aquaculture sector, potential areas of benefit were listed as the eradication of disease and pollution, and the reduction of labour costs. The China Fisheries Association has already signed a deal that will allow it to pull in data from its 1,000 member companies (which include manufacturers, processors, and distributors) in order to inform strategic decision making and drive early warning systems. The starting pistol for the global aquaculture race has, indeed, been well and truly fired.

References

Anderson, D, Hoagland, P, Kaoru, Y and White, A (2000) *Estimated Annual Economic Impacts from Harmful Algal Blooms (HABs) in the United States*. www.whoi.edu/cms/files/Economics_report_18564_23050.pdf (archived at https://perma.cc/B7EF-C5SZ)

Cabico, G (2018) SWS: 3.6m Filipino families experienced hunger in Q4 2017. www.philstar.com/headlines/2018/01/22/1780284/sws-36m-filipino-families-experienced-hunger-q4-2017#K8Mjqi67SxVacpTd.99 (archived at https://perma.cc/UD8R-3JY9)

Carney, B (2018) Sea Grant announces 2018 aquaculture research awards. https://seagrant.noaa.gov/News/Article/ArtMID/1660/ArticleID/2700/Sea-Grant-Announces-2018-Aquaculture-Research-Awards (archived at https://perma.cc/U4WT-R7MW)

Compton, L (2018) Huon confirms 120,000 salmon escaped in 'exceptional' May storms. www.abc.net.au/news/2018-09-12/huon-aquaculture-salmon-death-revealed-amid-transparency-calls/10230846 (archived at https://perma.cc/8C6C-54CS)

Dao, T (2018) Vietnam poised to become top player in ocean aquaculture. www.seafoodsource.com/news/aquaculture/vietnam-poised-to-become-top-player-in-ocean-aquaculture (archived at https://perma.cc/9BNY-65A8)

Fletcher, R (2017) Remote controlled aquaculture project launched. https://thefishsite.com/articles/remote-controlled-aquaculture-project-launched (archived at https://perma.cc/Y2F2-YVBX)

Fletcher, R (2018) Fish and chips: How computer analytics can transform aquaculture. https://thefishsite.com/articles/fish-and-chips-how-computer-analytics-can-transform-aquaculture (archived at https://perma.cc/3KQD-7VTX)

Food and Agriculture Organization (2018) Is the planet approaching 'peak fish'? Not so fast, study says. www.fao.org/news/story/en/item/1144274/icode/ (archived at https://perma.cc/P3GP-3CLU)

Garcés, J (2018) Algae fish death toll tops 4,000 tonnes in Chile. Fishfarmingexpert.com. www.fishfarmingexpert.com/article/algae-fish-death-toll-tops-4000-tonnes-in-chile/ (archived at https://perma.cc/SE46-DCDP)

Gatward, I, Parker, A, Billing, S and Black, K (2019) Scottish aquaculture: A view towards 2030: An innovation roadmap and sector needs study conducted by Imani Development and SRSL, on behalf of the Scottish Aquaculture Innovation Centre and Highlands and Islands Enterprise. https://pure.uhi.ac.uk/portal/files/2465192/Scottish_aquaculture_a_view_towards_2030.pdf (archived at https://perma.cc/A3XV-FQ3Q)

Glumac, T (2018) Hobart counts cost of damage as flooding clean-up continues. www.abc.net.au/news/2018-05-12/hobart-flooding-clean-up-worst-storm-in-decades/9753434 (archived at https://perma.cc/AUC5-QS9Z)

Godfrey, M (2019) China looks to big data to improve fisheries, aquaculture management. www.seafoodsource.com/news/supply-trade/china-looks-to-big-data-to-improve-fisheries-aquaculture-management (archived at https://perma.cc/L76C-HX93)

Gokkon, B (2018) Indonesian fish farmers get early-warning system for lake pollution. https://news.mongabay.com/2018/09/indonesian-fish-farmers-get-early-warning-system-for-lake-pollution/ (archived at https://perma.cc/98X5-NPCS)

Guardian (2016) Chile's salmon farms lose $800m as algal bloom kills millions of fish. www.theguardian.com/environment/2016/mar/10/chiles-salmon-farms-lose-800m-as-algal-bloom-kills-millions-of-fish (archived at https://perma.cc/7C5E-EAEV)

Hjul, J (2018) Tromsø in shock fish farm ban. www.fishupdate.com/tromso-in-shock-fish-farm-ban (archived at https://perma.cc/DS82-MYDT)

Hjul, J (2019a) Iceland aquaculture income up 34 per cent. www.fishupdate.com/iceland-aquaculture-income-up-34-per-cent (archived at https://perma.cc/QH5L-Q83U)

Hjul, J (2019b) Tromsø dilutes ban on sea farms. www.fishupdate.com/tromso-dilutes-ban-on-sea-farms/ (archived at https://perma.cc/J6GB-YLBT)

Karokaro, A (2018) Another mass fish kill hits Indonesia's largest lake. https://news.mongabay.com/2018/08/another-mass-fish-kill-hits-indonesias-largest-lake/ (archived at https://perma.cc/P65X-FDQ4)

Livescience.com (2009) Milestone: 50 percent of fish are now farmed. www.livescience.com/5682-milestone-50-percent-fish-farmed.html (archived at https://perma.cc/SY84-6VXL)

OECD (2016) *The Ocean Economy in 2030*. https://read.oecd-ilibrary.org/economics/the-ocean-economy-in-2030_9789264251724-en#page1 (archived at https://perma.cc/FRP6-QZZM)

Phys.org (2019) A new, escape-proof fish cage for a lice-free salmon farm. https://phys.org/news/2019-01-escape-proof-fish-cage-lice-free-salmon.html (archived at https://perma.cc/YVX6-MGTA)

Pobonline.com (2018) Satellite derived bathymetry aids fish farming. www.pobonline.com/articles/101284-satellite-derived-bathymetry-aids-fish-farming (archived at https://perma.cc/R5JK-DEQD)

Psa.gov.ph (2019) Fisheries situation report, January to December 2018. https://psa.gov.ph/fisheries-situationer (archived at https://perma.cc/6VMH-N7AZ)

PwC (2017) Sustainable growth towards 2050: PwC Seafood Barometer 2017. www.pwc.no/no/publikasjoner/pwc-seafood-barometer-2017.pdf (archived at https://perma.cc/VN5W-W7PK)

Salazar, M (2018) Latam eco review: Salmon escape, jungle drones, and a new biosphere reserve. https://news.mongabay.com/2018/09/latam-eco-review-salmon-escape-jungle-drones-and-a-new-biosphere-reserve/ (archived at https://perma.cc/8BVK-WSHC)

Shelleye.org (2017) ShellEye project. www.shelleye.org/About/For_Industry (archived at https://perma.cc/9PHN-6BJK)

Thefishsite.com (2018) Benefits of aquaculture are being 'overlooked'. https://thefishsite.com/articles/benefits-of-aquaculture-are-being-overlooked (archived at https://perma.cc/3RVS-VUTW)

Undercurrent News (2017) UK launches new aquaculture research center. www.undercurrentnews.com/2017/10/27/uk-launches-new-aquaculture-research-center/ (archived at https://perma.cc/LSF5-J3Q7)

Undercurrent News (2018) Norwegian net cleaning robot quick off the block. www.undercurrentnews.com/2018/09/20/norwegian-net-cleaning-robot-quick-off-the-block (archived at https://perma.cc/VUY7-FCSH)

US Soybean Export Council (2018) Aquaculture is Fastest growing food production sector, according to FAO report. https://ussec.org/aquaculture-fastest-growing-food-production-sector-fao-report/ (archived at https://perma.cc/U84Z-CLUP)

White, C (2018) Technavio report: Global aquaculture market's growth accelerating through 2022. www.seafoodsource.com/features/technavio-report-global-aquaculture-markets-growth-accelerating-through-2022 (archived at https://perma.cc/TYS6-CFVE)

World Bank (2014) Fish farms to produce nearly two thirds of global food fish supply by 2030, report shows. www.worldbank.org/en/news/press-release/2014/02/05/fish-farms-global-food-fish-supply-2030 (archived at https://perma.cc/49VD-AFJQ)

08

Hydrography and bathymetry

Bringing clarity to the ocean floor

In June 2016, the Nippon Foundation of Japan and the General Bathymetric Chart of the Oceans (GEBCO) made a bold, and very public, commitment. Speaking in Monaco at a specially convened gathering of over 150 scientists, scholars and ocean-reliant business stakeholders, they revealed the Seabed 2030 project, whose one objective was no less than the comprehensive mapping of the entire ocean floor by the year 2030.

GEBCO is a partnership project jointly supported by the Intergovernmental Oceanographic Commission of UNESCO and the International Hydrographic Organization, and – since its inception in 1903 – it has been mandated to map the ocean floor.

The boldness of the Seabed 2030 endeavour can only be fully comprehended when the current baseline is understood. It may come as a surprise to learn that the world's oceans are incredibly poorly mapped (charted). As we know, 71 per cent of the Earth's surface is covered with water, yet more than 85 per cent of the seafloor has yet to be surveyed to modern standards and full insonification of the seabed, according to the International Hydrographic Organization.

Staggeringly, 50 per cent of the world's coastal waters that merchant shipping routinely use remain uncharted. Amazingly, many important navigational charts are based on information that is well over 100 years old. The issue is not entirely related to available resources, as even the relatively wealthy USA has only mapped 30 per cent of its waters.

As many advocates in the field are keen to point out, to underline the size of the challenge (and highlight that it ought to be possible to achieve if enough willpower can be applied!), maps of both the Moon and Mars have been completed to significantly higher resolution than most of the Earth's seas and oceans.

Opened by Prince Albert II of Monaco (coincidentally the great-great-grandson of Prince Albert I, who founded GEBCO in 1903), the 2016 Forum for Future Ocean Floor Mapping intended to raise the bar on efforts to address this problem. Over the five days of the event it set out to convene and inspire important stakeholders to start taking action. Moving between panels, workshops and breakout sessions held in the Monaco Oceanographic Museum, the Monaco Yacht Club and the Novotel Monte Carlo Hotel, expert delegates grappled with the size of the challenge, how to take the ambition to the rest of the ocean community and – crucially – explored what kind of activities and areas of focus would allow the ambition to be achieved in the decade and a half to follow.

At the end of the conference, all delegates endorsed the Seabed 2030 core objective, and focused in the first instance on the sharing of existing bathymetric information to build a comprehensive and up-to-date baseline picture. In addition, delegates highlighted the greater need for coastal and developing nations to have access to the latest tools and technology to enable them to understand in detail the makeup of the seabeds surrounding their nations.

The role that technology can play in the Seabed 2030 objective is obviously critical. It is clear that if traditional methods were able to meet this challenge, more progress would have been made over the course of the past century. The ambition can only be satisfied by utilizing the latest innovative measures, which we will explore later in the chapter.

However, with no shortage of resource demands to tackle Blue Economy issues, how urgent an imperative should mapping the seabed be? Despite still being in the dark on comprehensive details of 85 per cent of the oceans depths, what is the real value of seafloor mapping? Can it be considered a priority?

One data set, multiple uses

The ocean bed is constantly changing, influenced by tectonic shifts, seismic activity, marine biodiversity, weather and currents such that many charts have inadequate or sparse information, and there is – as GEBCO underlines – a real thirst for as much new and modern bathymetric data as possible.

It is worth being clear about definitions. The terms hydrography and bathymetry can be confused by those with only a passing interest. Bathymetry is often described as the underwater equivalent of topography. Whereas the latter is the study of the shape and features of surfaces on land (including natural formations such as mountains, rivers, lakes and valleys), bathymetry is the same underwater – focusing on establishing the contours and features of the subsea terrain. Where topography depicts the shape of the land through contour lines or differentiated colours, bathymetry represents depth variations in the seafloor, which can be visualized through colour coded depth contours.

Bathymetry is a subset of the wider discipline of hydrography, which also includes a focus on shoreline features and shape, and broader ocean characteristics such as those that relate to tides, currents and waves. Hydrography also concerns itself with the physical and chemical through water column properties. These include temperature, salinity and sound velocity.

These two related disciplines are far from only being of interest to academics alone. Detailed knowledge of global bathymetry is critical for understanding how Earth's systems interact and to support everything from coastal zone management to environmental protection; from tsunami modelling to inundation forecasting – especially important for the Small Island Developing States known to be most vulnerable to natural disasters.

The business drivers tell a similar story of need and associated growth. Business intelligence company Markets and Markets predicts that the global hydrographic survey equipment market is expected to grow to £2.08 billion by 2022, at a compound annual growth rate of 5.39 per cent during the forecast period (2017–22) (Marketsandmarkets.com, 2018). The company assigns this expected growth to the rise in maritime commerce, the new complexities of coastal zone management and a general increase in demand for modern detailed nautical charts.

Actionable intelligence

Identification of the largest market areas for hydrography services begins to highlight the breadth and importance of the sector. Offshore oil and gas is the largest market segment. Stakeholders within this sector require detailed hydrographic services to enable them to plan and install offshore production platforms and pipelines, while safely managing the ongoing inspection and maintenance thereof.

Hydrographic surveys provide the bathymetry and object detection data to produce navigational and other charts for safe vessel transit and seafloor exploration activities. Clients here include governments and their respective regulatory and safety bodies, who always need to keep up to date with the ever-shifting underwater landscape. Coastal nation states are legally bound to chart and provide navigation safety services for their own waters.

As an example, in the summer of 2018 the UK Hydrographic Office announced a new bathymetry-based project conducted in partnership with the Cayman Islands, a British Overseas Territory in the western Caribbean Sea. Funded by the UK's Conflict Stability and Security Fund (which had an allocated budget of £1.28 million for the financial year 2018–19), the project aimed to support the Cayman Islands to comply with its obligations to the Safety of Life at Sea Convention by supporting safe navigation in the country's waters.

Areas to the west and south of Grand Cayman were surveyed, as well as the south-western tip of Cayman Brac. Two surveying techniques were employed to capture the data afresh. First of all, some areas were surveyed using airborne Lidar. With this approach, water-penetrating laser radars are attached to fixed-wing or rotary wing aircraft aeroplanes and flown across the survey area to collect underwater topography. The high-powered lasers transmit electromagnetic energy (specifically, near-infrared and green), making highly accurate time-difference measurements that allow calculation of the seafloor depth. Logistics reasons mean that Lidar is mainly used in coastal areas only, and – as results degrade as the turbidity (murkiness) of water increases – is often limited to depths of less than 50 metres.

A capability in development since the 1960s, the Cayman Lidar surveys were complemented by a technology that started to gain traction a decade later. Multi-beam echo-sounders (MBES) are now the industry standard for bathymetric data gathering in water deeper than 5 metres. Replacing single-beam echo sounders, MBES were feted for representing the most revolutionary advancement in seafloor mapping since the advent of the lead line several thousand years ago.

MBES systems – as their name suggests – are a type of sonar that emit multiple sound waves into the water under a ship, and determine water depth by recording the time it takes the sound waves to bounce off the seafloor and return to the receiver. It effectively blankets the seafloor with pulses of sound energy (a fan-shaped array of pulsed energy divided into a series of beams) that then provide a range of measured depths. MBES changed the game as the system was able to collect a much wider swathe of seafloor information within the same timeframe, and ensure that there were no gaps in data, as long as the vessel is accurately positioned and surface noise does not interfere with operations.

Ports, boundaries and offshore developments

Moving on from the specifics of the Cayman Islands project, there is also a great need to map properly the world's commercial ports and harbours, especially the 2,000 largest in the world. This relates both to regular surveying of existing ports to ensure that the dredged channels and maintained depths beside berths or jetties have not reduced through silting or undergone significant change, as well as to helping plan and build new ports, such as the one to be delivered in southern Bangladesh's Barisal Division. Officially inaugurated in 2016, following an establishing Act of Parliament in 2013, Payra Deep Sea Port is the developing country's third and largest sea port, built at total cost of somewhere between £8.4 and £11.4 billion. Such investment is intended to provide a significant return – in this case allowing the country to accommodate much larger sea-going vessels directly, rather than relying on transhipments through regional hubs such as Colombo or Singapore. High-resolution bathymetry is in many ways the cornerstone of such development.

From a legal and safety of life at sea perspective, there is an ongoing need for hydrography to help define and refine Exclusive Economic Zones. Such activity helps to establish nations' rights over seabed 'real estate' and, unsurprisingly, the associated oil, gas and mineral rights that pertain to such boundaries.

These more traditional needs have in recent years been bolstered by the demands of two of the Blue Economy's key growth sectors. Hydrography is becoming increasingly important for the identification and assessment of locations suitable for offshore wind farms; and – as aquaculture has emerged as an increasingly important sector for the provision of food and protein – its growth has required modern coastal and deep water surveys for infrastructure installation.

So, there is no doubt that the environmental and commercial pull for hydrography looks set to increase in the coming years. However, the industry standard tools – as reliable as they are – look incapable of driving the kind of step change in seabed mapping demanded by the Seabed 2030 call to arms.

We have seen that Lidar is limited to depths of less than 50 metres, and more often than not can only reasonably operate within a 350-mile range of a coastal airfield. While MBES was indeed a truly disruptive innovation when it came to market, these powerful systems are still limited by their reliance on existing ocean-going survey vessels, and the associated limits (flexibility of routes on joint survey missions) or costs (especially manpower for survey vessels) of traditional ocean-going vessels.

The success of the GEBCO initiative, then, may rest on the further development of two innovative technologies that appear prominently in many of our Blue Economy sectors: satellites and autonomous vessels.

Thankfully, both the use of satellite-derived bathymetry and the development of hydrographic systems on autonomous or remotely operated vessels have been making great gains in the past few years. And the market is very much starting to see the benefits.

From theory to action

In January 2018, unmanned vessel company ASV Global (at the time an independent company but since acquired by New York-based L3 Technologies) deployed its C-Worker 7 autonomous vessel to support a pipeline operation led by Subsea 7 off the Egyptian coast. Utilizing both a multi-beam echo sounder and side-scan sonar, the 7 metre unit supported the main pipe-laying cable ship by providing autonomous touchdown monitoring capabilities – to detect unexpected obstacles before they compromised the integrity of the pipe-laying process. Traditionally, this monitoring is carried out by a remotely operated vehicle supported by a dedicated support vessel, which increases the costs of the operation significantly.

Operating over 37 days, the C-Worker 7 removed the need for the expensive additional support boat, thus significantly reducing costs. While actual figures were not available, additional manned touchdown support vessels can represent up to 90 per cent of the monitoring cost (especially in deep waters where dynamic position can be essential), so there is a good margin to be eaten into. Cost savings aside, at the time the company was keen to highlight that its autonomous operation also removed any potential risk to personnel working on the project, as they did not need to go to sea for this purpose.

Then in May 2018, Florida-based company SeaRobotics delivered the final two (of four) autonomous, unmanned surface vessels to the Canadian Hydrographic Service, a division of Fisheries and Oceans Canada. The vessels is 2.5 metres long and packed with sensors, the most important of which is the multi-beam echo sounder, provided by Texan company R2Sonic. The units will be used to deliver highly accurate bathymetric and hydrographic surveys

In the summer of 2018, the port of Antwerp announced that it would for the first time be trialling the use of a fully autonomous survey boat. The *Echodrone* is powered by electricity and is able to undertake echo sounding – without human involvement – to provide up-to-date seabed inspection data to inform understanding of which areas of the port may need dredging in order to maintain safety of navigation. Working within a very busy port environment has driven one of the system's key innovations, in that the *Echodrone* uses verified navigation data in the cloud to enable it to move safely around the port. This system means that the *Echodrone* – much smaller and nimbler than its larger, manned sister survey vessel – can continue to operate safely in heavy shipping traffic.

Order from chaos

Hydrographic surveys can be most urgently required after major environmental events. Hurricanes and tsunamis can significantly alter the features of the seabed, making it almost impossible to know if traditional marine passages are still safe to navigate, much less understand how coastal areas may be newly susceptible to storm surge. This was certainly the case following the devastation caused by Hurricane Florence, which tore through the coastal states of North and South Carolina in September 2018. When a severe weather event passes over a coastal country often the airports are immobilized and the quickest way to get aid in and get the economy running again is through access from the sea to ports and harbours

The National Oceanographic and Atmospheric Administration (NOAA), part of the US Government's Department of Commerce, was quick to act. Its Office of Coast Survey navigation response team deployed a range of capabilities in order to gain an up-to-date picture of the seafloor. Most notably, they surveyed Murrells Inlet, South Carolina, using their autonomous surface vessel the *EchoBoat*, which was equipped with both a side-scan sonar and a multibeam echo sounder, and the REMUS 100 autonomous underwater vehicle.

We can therefore see that autonomous hydrographic options are starting to be utilized in a wide range of ocean settings: by port authorities, governments and related agencies looking to ensure safe passage – as part of normal business or in response to major environmental events – and by cable companies looking to support the laying of new subsea infrastructure.

So why are such businesses, authorities and agencies jumping on autonomous vessels to deliver hydrographic data? For many reasons.

Currently, manned survey vessels are sent on pre-defined missions to gather data, though they do of course have the flexibility to re-task based on what they find while surveying. Data is often processed in near real-time by the surveyors, integrated into a geographical information system (GIS), ranging from GIS representations from the nautical chart to the representation of bathymetry as shaded relief or as a digital elevation model. However, this is all a relatively expensive and logistically difficult endeavour.

Surveys need to be booked several months – if not years – in advance. Operating in harsh conditions can be troublesome; finding crews to support midwinter expeditions can be difficult; and there is always risk associated with sending humans into challenging conditions.

Installing MBES or other hydrographic equipment on unmanned surface vessels – either running autonomously or tasked remotely using space-based assets – has the potential to reduce costs and broaden reach significantly. Such systems could bring new and currently unavailable benefits to market, and therefore promise much-needed affordable additional global capacity, by unlocking greater affordability (lower capital and operating costs); rapid deployment and recovery, vastly reducing logistics; enhanced safety, by providing the ability to operate in harsh conditions without threat to human life, as the need for crew or divers in the water is removed; and fast access to the data gathered and remote, real-time control of the vessel.

While great progress has been made in this field, scalability of autonomous hydrographic solutions still relies on a number of additional technology developments. Autonomous vessels may need to increase the speed at which they can sail in order to create truly eye-catching service offers. Onboard power options need to progress further to ensure true stability of operations. And the final hurdle relates to associated costs. Autonomous vessels rely on satellite connectivity for connection to the vessel for command and control from shore, and connection to the vessel for transmitting bathymetric and other hydrographic data from the vessel to shore, to information-hungry customers and scientists. This can be expensive, but some great work is already being undertaken in this respect in the development of systems that process raw data on board the vessels, so that much lower bandwidths are required to transmit the processed data.

Despite these challenges, many industry figures are very positive about the potential of autonomous vessels within the fields of hydrography and bathymetry, with one scientist telling us: 'You can pay a heavy price for hydrographic mapping. Even though most of it is in shallow water, it comes with a big price tag. Unmanned surface vessels can undoubtedly be a force multiplier as they can provide much longer operation at reduced cost.'

Another hydrographer reinforced the value of true autonomy when they said: 'Sending out crews in dirty weather in the middle of winter is never attractive – and it's expensive – so being able to deploy an unmanned surface vessel for that sort of mission is interesting on all sorts of levels.'

But while this innovative technology begins to prove itself further within the market place, another is also emerging that removes the need to deploy an ocean-going vessel at all.

Safety blanket

As we have seen, coastal communities face a continuous battle to map their seaspace so that they can understand changes in the subsea landscape and react accordingly. Statistics show that over 70 per cent of all natural disasters experienced in coastal areas all over the world are related to extreme climate events. Climate change is also anticipated to increase the coastal hazard threat trends (in both intensity and frequency) through sea-level rise, floods and storm.

These facts align to the knowledge that coastal populations are dramatically increasing (leading towards more demographic vulnerabilities), with African coastal nations, parts of Asia and Southeast Asia and Small Island Developing States (SIDS) all needing to protect clusters of the most vulnerable coastal communities. SIDS in particular face grave risks from hurricanes, tsunamis and sea-level rise. This is exacerbated by the fact that they also depend heavily on fisheries for both food and income. Therefore, when their fisheries are affected – whether from natural hazards or ineffective management — the barriers to recovery are that much more significant.

This range of challenges is nowhere more prevalent than in the Caribbean region. Over 300 natural disasters have hit the region since 1950, with approximately 250,000 people losing their lives and more than 24 million affected through injury, death or loss of homes and livelihoods. The International Monetary Fund considers many Caribbean islands to be among the 25 most vulnerable nations in terms of disasters per capita or land area, with impacts potentially catastrophic to livelihoods. When Hurricane Maria hit Dominica in September 2017, for example, the cost of the damage amounted to more than 200 per cent of the country's gross domestic product.

Focusing specifically on the largest tourist cities on the Mexican Caribbean coast alone, academic research published in January 2019 predicted that the economic impacts of increasing sea-level rise could range anywhere from £250 million to £1.75 billion (should sea levels rise 3 metres and no mitigating actions be taken) (Ruiz-Ramírez et al, 2019). As the report authors declaim, the scale of this potential impact underlines the need to explore in greater detail potential future impacts – be they economic or ecological – of climate variability and change on the Mexican Caribbean coastline.

'Irmageddon'

As well as the longer-term need for hydrographic data, as the earlier NOAA example in the North and South Carolinas highlights, speedy data collection is most needed following a major environmental event. This was brought into sharp focus in the Caribbean in September 2017, when category 5 Hurricane Irma devastated large parts of the Florida Keys and the north-eastern Caribbean area. It became the second-costliest Caribbean hurricane on record, with the phrase 'Irmageddon' attempting to highlight its impact.

Amongst this carnage, one person died in Barbuda, and three people lost their lives in Antigua. Barbuda – covering only 62 square miles – was the first to be hit by Irma. Battered by 185 mile winds, and 8 foot storm surges, over 90 per cent of the island's property was affected, with the repair bill estimated at approximately £150 million. All 1,800 shell-shocked residents were evacuated two days later as the subsequent threat of Hurricane Jose rolled onto the horizon; as of February 2019 around 500 of the island's citizens still had not returned.

Against this backdrop of devastation, then, and with the continued threat of new climate events apparent, it is understandable that calls were made to map the country's coastal area, to understand what may have changed and what measures may be needed to better prepare ongoing coastal defences.

The Government of Antigua and Barbuda needed to be able to compare offshore water depths pre- and post-Irma, and were supported in this endeavour by the UK's Centre for Environment, Fisheries and Aquaculture Science, who commissioned a survey under the auspices of the Commonwealth Marine Economies Programme, a UK Government programme that aims to support SIDS and help them to grow sustainably.

The resulting data sets helped digital change detection analyses to ascertain seafloor changes as a result of Irma, to contribute to the country's national adaption strategy. The combination of onshore and offshore analysis in particular aids impact modelling related to coastal flooding.

The remarkable aspect of the survey, however, was that no marine vessels were involved, as the entire marine data collection operation was conducted from space. Specialist company TCarta won the public procurement exercise to deliver the project using satellite-derived bathymetry (SDB), a technique that is particularly useful in the kind of clear, shallow waters found in abundance in the Caribbean.

Whereas deep seas hydrographic tasks present their own logistical and safety challenges, shallow water bathymetry is no less challenging. Shallow areas can be difficult to access by boat, and staff and equipment can be susceptible to ocean currents, rendering traditional hydrographic survey techniques (eg using echosounders on vessels) unsuitable. There are no such issues with SDB, which operates in some way on the same principle as echo sounders, in that it measures the distance of reflected bands of energy except that the SDB process uses light instead of sound, and does so using sensors on Earth observation satellites (supported by computational algorithms) rather than attached to the hull of a survey ship. In particular, the system utilizes visible light (eg red, green and blue), sometimes supplemented by other, non-visible, reflected and emitted bands such as near and far infrared, which have longer wavelengths than visible light.

A light in the dark

Such inherent reliance on light means that one of the main limiting factors of the process relates to the clarity of the water being surveyed – in essence, how clearly and accurately you can see the ocean floor from space. This means that the optimal depth for SDB is in the 20–30 metre range, but even then, some additional factors need to be taken into account. The results for coastal areas that suffer from significant algal bloom (microscopic algae or bacteria in the water that can lead to a coloured scum on the surface), lots of sediment in the water column or vigorous wave action, may be limited.

One of the earliest signifiers of the potential power of the approach came in 2013 when German company EOMAP (a spinout from the German Aerospace Centre) worked with the James Cook University to demonstrate its SDB capabilities. It did so by mapping approximately 350,000 square kilometres of Australia's Great Barrier Reef, to a horizontal resolution of 30 metres. This was the world's first digital map of the world heritage-listed area, as up until that time, nearly half of the shallow water area of the reef area had not been mapped using modern digital surveys due to problems of accessibility and navigability.

TCarta were awarded the Antigua and Barbuda contract on 20 December 2017. As a pre- and post-hurricane comparison was required, their team originally dug back into their archive of satellite imagery to see if a suitable baseline was present, and were relived to find several suitable images collected prior to Hurricane Irma.

They then looked at what had already been captured by satellite since the hurricane activity in early September. They located suitable images over Barbuda at that time, but found nothing that they could use across the larger island of Antigua, so they commissioned Digital Globe – who own and operate a constellation of high-resolution Earth observation satellites – to capture new imagery to fill in the gaps.

By 28 December – eight days after the award of the contract – TCarta had assembled a full set of useful and appropriate images that provided seamless land and shallow water marine surface, both before and after Hurricane Irma. The project was completed entirely remotely, which meant that no mobilization or permits were required – itself an issue when the true nature of the seafloor is obviously not known. As well as the astonishing speed of the initial data capture (some additional images were added later), TCarta estimated that the cost was about 10 per cent of the amount for traditional survey methods of boat or aircraft.

One of the more immediately apparent findings was that the hurricane had led to a sediment shift in one of the main known navigation channels around the island. By understanding the details of this change, the port authority could begin to manage better the shipping lane, and prioritize where any corrective engineering work might be needed. For an island economy already on its knees, enabling ships to get in and out of port is absolutely critical for the infrastructure to start improving and redeveloping. With this layer of intelligence the authorities are able to focus their efforts and understand what needs to be repaired in order to start getting the economy back up and running.

Focusing on the environmental domain, the TCarta team noticed significant changes in sediment dispersal in Fellingo Shoal, off the northern coast of Antigua. Analysts concluded that the shoal was effectively being 'squashed', and pushed into a new area. Understanding such sediment instability helps ecosystem services, for example, to provide intelligence to the fishing industry, as significant sediment shifts can obviously affect fish patterns.

This project alone proves that satellite-derived bathymetry offers significant gains to the world of hydrography, providing both cost-effective and speedy analysis of a given area, two factors of magnified importance in disaster areas.

The other major advantage of SDB over traditional methods is that analysts can access historical data. The real value in the Antigua and Barbuda project was in being able to see the change in the coastal area. If only traditional approaches were available, it would obviously not be possible to go back in time to send out a survey boat or plane. However, the Digital Globe Earth observation satellites had recently passed over the area and collected relevant data, which provided the project's baseline comparison.

Setting the future course

The leading lights behind the launch of Seabed 2030 have refined their approach over subsequent gatherings since their initial exploratory event, including celebrating the ambition to a much wider audience at the first ever UN Ocean Conference in New York in 2017. More detailed plans have been developed, based around three pillars of activity.

The first focuses on data discovery, to ensure that all existing bathymetric data is safely stored in national archives. It is considered that such data may have been collected on sub-national projects whose project leads may now be willing to offer them up to aid the Seabed 2030 project. To support this element, globally distributed Regional Data Assembly and Coordination Centers will be established, to help identify existing data from their assigned regions that are not currently in publicly available databases and seek to make these data available. Cutting across this work (as in so many collaborative and open data projects) is the need to develop protocols for data collection (including resolution goals), and common data standards and software approaches. A Global Data Assembly and Coordination Center will co-ordinate this work, as well as integrate the output of the regional grids.

Building on this step, the second pillar of activity aims to create high-resolution seafloor maps to share with the public, via the Seabed 2030 website, Google Earth and other platforms such as the Ocean Basemap service provided by GIS specialists Esri.

This pillar received a boost in September 2018 when – after a period of testing – NOAA enabled public access to a new crowd-sourced bathymetry database that collated information drawn from bathymetric observations provided by organized crowdsourcing programmes and citizen scientist volunteers. At the time of launch, the database already had more than 117 million points of depth data.

With the third and final pillar comes the mammoth task of filling in the gaps in the global bathymetric data set. Aside from a renewed call to crowdsource bathymetry data sets from fishing vessels and recreational vessels, much hope lies in the potential of new technologies as force multipliers to enable the completion of a high-resolution map of the world's seabed by the year 2030.

The call is starting to be heard. In June 2018, Texas-based surveying company Ocean Infinity donated 120,000 square kilometres of data to the Seabed 2030 project. This data was collected by a cutting-edge fleet of eight autonomous underwater vehicles, which were actually deployed in the sadly unsuccessful search for the missing Malaysian airliner MH370. Offshore surveying company Fugro have also donated 100,000 square kilometres of transit data, while the Australian government provided 710,000 square kilometres. The willingness of all of these stakeholders to provide their data for such an important secondary use bodes well for the headline aim.

Conclusion

The final comment on emerging technologies in this domain is one that relates to many changes and advancements in Blue Economy sectors, and that is about the potential complementarity of such new capabilities. Many technologists and entrepreneurs can too easily reach for the 'disruptive' adjective to describe their new offer. They will 'disrupt' and replace traditional approaches with their better, smarter, cheaper, more reliable systems, quickly relegating extant options to the scrapheap of redundant technologies. In practice, this is very rarely the case, at least not within a short timescale.

What more often happens is that the new technology does indeed provide quality, capacity or cost-saving benefits, but in ways that find their place *alongside* existing approaches, rather than replacing them.

While owners of MBES-enabled autonomous vessels may promise that in time they will replace entirely more traditional methods, a blended outcome is more likely. Such vessels could work in tandem with more expensive and complex offers (eg survey boat or aeroplane) to increase effectiveness and overall coverage. Hydrographic scientists have expressed active interest in deploying autonomous vessels as part of wider planned survey missions. This could mean sending autonomous vessels out ahead of the main mission to conduct initial surveying that will help to focus subsequent energies, or to leave them behind to undertake additional survey work in areas of interest when the crews of the time-limited main survey vessel have to return home.

In similar fashion, for all its promised speed and cost enhancements, satellite-derived bathymetry should not be seen as sounding the death knell of all other traditional forms of bathymetry. First of all, SDB algorithms require some registration with known, validated depth points, which provide registration points and a frame of reference to help calibrate the space-based system. Other existing data is also useful to power accuracy within SDB projects. Not all sea bottom types are the same, as sand, coral and seagrass can have varying reflective signatures, threatening to disrupt the SDB algorithm's accuracy unless such data is considered. SDB is most effective in clear waters, but its use is limited in areas where waters are turbid with particles (silt) suspended in the water column.

Moreover, like other autonomous options, SDB can be seen as just one more new option within the hydrography toolbox. In 2017, for example, EOMAP were commissioned by Land Information New Zealand to undertake a satellite-derived bathymetry survey of the Tongan archipelago and surrounding areas. As well as providing invaluable new shallow-water data, it was acknowledged from that start that the operation had an additional strategic aim of pinpointing where additional, traditional survey methods should follow. The SDB component was therefore used as a cost-saving technology in its own rights, as well as a targeted reconnaissance tool that helped to further optimize the planning of traditional surveys.

Traditional survey techniques are by their very nature expensive (ships and people) and generally slower to complete. With the advent of technological advancements through remote sensing and autonomous systems, the choice and flexibility has increased markedly, and with that, enabling agility. That said, rarely is there one system which provides the optimal solution and one size does not usually fit all. Analysis of the survey task required balanced against environmental factors and physical factors of where the survey is to take place will determine the optimum solution. While it could be through one system or sensor, more often it will in fact require a hybrid or blend of differing hydrographic collection capabilities. Therefore, agile and effective utilization will in its very nature enable cost effectiveness.

This 'system of systems' collaborative approach offers perhaps the greatest comfort to the passionate hydrographers behind the GEBCO Seabed 2030 initiative. While many parts of the seabed are still a relative mystery, the new tools available undoubtedly provide a much-needed boost to a century-old Blue Economy ambition whose time may finally be coming.

References

Marketsandmarkets.com (2018) Hydrographic survey equipment market by type and region: Global forecast 2022. www.marketsandmarkets.com/Market-Reports/hydrographic-survey-equipment-market-38915154.html (archived at https://perma.cc/7EP8-RZR8)

Ruiz-Ramírez, J, Euán-Ávila, J and Rivera-Monroy, V (2019). Vulnerability of coastal resort cities to mean sea level rise in the Mexican Caribbean. *Coastal Management*, 47, pp 23–43. www.tandfonline.com/doi/abs/10.1080/08920753.2019.1525260?af=R&journalCode=ucmg20 (archived at https://perma.cc/B2QS-8ZGZ)

09

Ocean conservation

Protecting and preserving the future of the seas and oceans

The visibly emaciated whale that washed up on the coast of Mabini town, 80 miles south of the Philippine capital Manila, was harbouring a deadly secret, and the surprising cause of its own demise. The 15-foot Cuvier's beaked whale, a species that can be found in waters as far flung as the Aleutian Islands off Alaska, and as far south as New Zealand and Tierra Del Fuego, was examined jointly by scientists from the Government's Bureau of Fisheries and Aquatic Resources and the local D' Bone Collector Museum Inc in Davao City (Ellis-Petersen, 2019). They concluded that the great beast had died from starvation, but the underlying reason for that was something of a shock even to hardened marine observers. It was unable to eat because its stomach was full of 40 kilograms of plastic that it could not pass nor break down, the tragic haul consisting of everything from grocery bags to rice sacks.

It was a difficult end. Emaciated and too weak to swim on its own, the whale became stranded in shallow waters, dehydrated, and was seen to vomit blood as it eventually succumbed to its fate. 'Disgusting and heart-breaking,' was the succinct analysis of one of the investigators.

This unfortunate event in March 2019 was sadly just the latest in the ever-more regular caseload of marine animals being ravaged by the seemingly rampant problem of ocean plastic pollution. A dead sperm whale picked up in Wakatobi National Park in neighbouring

Indonesia in November 2018 was found to have 6 kilograms of plastic waste in its stomach, comprised of 115 cups, plastic bottles, bags, sandals, and a sack containing more than 1,000 pieces of string (New Straits Times, 2018). Five months earlier, a dead pilot whale was discovered on the marine border between Thailand and Malaysia with its stomach also clogged up with 80 pieces of plastic rubbish (Zachos, 2018).

In September 2018 in the United Arab Emirates, Sharjah's Environment and Protected Areas Authority's Scientific Research Department and Breeding Centre for Endangered Arabian Wildlife released worrying research. The project's researchers had examined 14 dead green sea turtles, all of which had washed up in the Gulf of Oman. Eighty-six per cent of the turtles had ingested marine debris, mainly plastics (Burgess, 2018). Again, the variety of the ingested debris was eye-watering as the turtles were found to have swallowed ropes, fabric, cotton buds, woven and regular plastic bags, fishing lines, hooks, nets and traps. Researchers noted that the majority of these items were white or transparent, and suggested that sea turtles may eat them as they mistake such plastic for similarly hued jellyfish. The green turtle (Chelonia mydas) is already endangered in these waters, so a new man-made threat is particularly unwelcome.

The ocean plastic pollution problem has increased exponentially in recent years, and may still get worse before it gets better. Once in the sea, plastic debris can end up in several places – washed up again on beaches, floating on the surface, suspended in the water column or ingested by all manner of marine animals, and the burgeoning body of academic research addressing related issues in all parts of the ocean makes for grim reading. To reference just a few, in September 2018 the British Antarctic Survey revealed that the amount of plastic washing up on the shores of remote South Atlantic islands had increased by a factor of ten in the past decade, for the first time approaching levels seen in industrialized North Atlantic coasts (Bas.ac.uk, 2018).

The following month, Greenpeace reported that 17.5 million pieces of plastic waste are flushed from Hong Kong's Shing Mun River into the sea each year. During wet weather, it reported, this meant that 56 pieces of waste plastic flow from the river into the sea every *minute*.

The most abundant items recorded both by human observers and video camera were said to be food packaging, takeaway boxes, straws, tableware, bags, bottles and Styrofoam (Kao, 2018).

At almost the same time as the Greenpeace study, additional research published in the journal *Environmental Science and Technology* estimated that just 10 river systems are the source of up to 95 per cent of all of the plastic that finds its way into the ocean (The Maritime Executive, 2017). Eight of these major culprits are in Asia (the Ganges, Indus, Yellow, Yangtze, Hai He, Pearl, Mekong and Amur), with the final two (the Nile and the Niger) to be found Africa. Not surprisingly, the German researchers note that they are all located in densely populated areas where littering is a known problem.

One of the most comprehensive studies in this field was published in 2015 in the journal *Science*, by the University of Georgia. It estimated that approximately 8 million metric tons of mismanaged plastic waste entered the ocean from the 192 countries with a coast bordering the Atlantic, Pacific and Indian oceans, Mediterranean and Black seas (UGA Today, 2015). This equalled just under 3 per cent of all plastic waste produced by these nations, with that plastic waste making up approximately 11 per cent of the headline total of 2.5 billion metric tons of solid waste produced. However, that research was based on 2010 data, so there is a strong argument to say that the issue will have increased significantly since then.

A broader challenge

Aside from issues related to individual marine animals suffering in various ways after ingesting plastic waste, this pollution phenomenon threatens to provide much broader challenge to ocean ecosystems. In 2017, for example, scientists were surprised to record that entire communities of coastal species had been transported thousands of miles between Japan and North America on what amounted to plastic debris rafts. Nearly two-thirds of the 289 living species discovered (which included molluscs, worms, bryozoans (tiny marine invertebrates) and crustaceans) had never been seen on the US West Coast before; they had completed their journeys on debris, much of it plastic, that had been swept out to sea in the aftermath of the 2011 earthquake and tsunami in Japan (Wyatt, 2018). While an unusual event occurring after the ocean abnormalities of the major tidal wave, scientists noted that such journeys were made possible as their 'vessels' were made of fiberglass or other plastic materials that do not decompose and were said to have been easily able to survive six or more years at sea. Although such storm-surge related incidents will not take place often, the mass of plastic waste in the ocean clearly has the potential to change some quite fundamental principles of how oceans function. What, for example, might happen if more invasive marine species are carried into new, non-native waters on this flotilla of ubiquitous plastic?

So what can be done? The most visible sea-borne example was launched in September 2018, with the development and deployment of the Ocean Cleanup System (The Ocean Cleanup, 2018). The Great Pacific Gyre (or Great Pacific Garbage Patch) is a vast rotating ocean vortex located 1,200 nautical miles off the coast of San Francisco that traps surface material at its centre, thus concentrating any debris as a very visible symptom of the ocean plastic pollution issue. It is estimated to contain 1.8 trillion pieces of plastic that weigh 80,000 tonnes, and is sadly not the only such gyre. Similar phenomena exist in the Indian Ocean; the north and south Atlantic; and the north and south Pacific.

Designed in the Netherlands by Dutch inventor and project figurehead Boyan Slat, the team grew to more than 70 engineers, researchers, computational modellers and other assorted scientists. The Ocean Cleanup System aims to collect as much floating waste as possible at sea by creating 'a coastline where there isn't one'. It harnesses a 600 metre floating device that carries a tapered 3 metre skirt underwater that is designed both to trap large volumes of plastic in its horseshoe shape and allow marine life to continue to flow freely.

Launching from Alameda, California, the system was towed the required 1,200 nautical miles off the San Francisco coast by project partner, shipping company MAERSK. One of UK company AutoNaut Ltd's unmanned surface vessels also accompanied the mission. As well as live streaming video footage above and below the water, the AutoNaut collected environmental data (including on currents, water quality, and wave height and direction) autonomously and utilized passive acoustic monitoring sensors to detect and track whales and dolphins so that they would not be disturbed by the innovative system.

The project team's ambition is nothing less than reducing the Great Pacific Garbage Patch by 50 per cent in five years. However, its first major deployment did not end successfully. Plastic spilled out of the boom system. Currents drove plastic on the outside rather than inside the system. After collecting 500 kilograms of plastic over a three-month period, a 60-foot section of the device detached from the rest of the system. This was caused, according to the team, by 'material fatigue' (Marine Insight, 2019).

Some researchers had publicly expressed strong scepticism that the project would work, citing the potential for biofouling as one of the main barriers to such a system enduring at sea. In the main this was not necessarily because of the principle of the concept, but borne of the understanding of the complexities of at-sea engineering projects of such grand scale. Some marine scientists did strongly suggest, though, that the project's system could actually harm the marine ecosystem as it chugged across the ocean (a claim refuted by the Ocean Cleanup management within their environmental impact assessment).

Micro waves

While the optimistic team behind the system re-group to make refinements, others are working equally as diligently to tackle an issue not possible to be addressed by the Great Ocean Cleanup. A well as the more obvious floating plastic debris, microplastics (categorized as small pieces of plastic less than five millimetres long) are also having a destructive effect in the world's seas and oceans. A problem that has only come to light relatively recently, academics are quickly catching up in quantifying how much of an issue microplastics are presenting at sea. The full extent of this problem is becoming understood by the amount found within marine creatures, who appear to be consuming it at an alarming rate.

A 2015 report in the scientific journal *Marine Pollution Bulletin* examined the digestive tracts of 263 fish across 26 species that had been caught off the Portuguese coast. Microplastics were found in 17 of the species, corresponding to nearly 20 per cent of the catch. The polymers identified included polypropylene, polyethylene, alkyd resin, rayon, polyester, nylon and acrylic (Neves et al, 2015).

In a February 2019 report published in the *Royal Society Open Science* journal, scientists reported that they had found, for the first time, that microplastics had been ingested by organisms in the Mariana trench and five other areas of the ocean with a depth of more than 6,000 metres. This led the research team to conclude that 'it is highly likely there are no marine ecosystems left that are not impacted by plastic pollution' (Watts, 2019).

Moreover, as microplastics can be as minute as 100 nanometres in diameter (a billionth of a metre), they can even be eaten by corals. Indeed, research published by Duke University in September 2017 in the online edition of the journal *Marine Pollution Bulletin* suggested that corals even have a discerning palate when it comes to microplastics (Applied Sciences from Technology Networks, 2017). The study suggests that they prefer 'clean' types of plastic (unfouled microplastics) by a three-to-one margin over 'dirty' ones (weathered microplastics fouled with a bacterial biofilm).

Before one conjures notions of plastic being some sort of subsea treat for corals, however, a major study published in *Science* in January 2018 revealed that plastic pollution increases the likelihood of disease in these important yet delicate organisms by a factor of over 20, from four per cent in areas with no plastic waste evident to nearly 90 per cent for corals that are in contact with plastics (Gabbatiss, 2018). The study also estimated that a total of 11.1 billion plastic items are ensnared on reefs in the Asia-Pacific region (from the ubiquitous shopping bags to disposable nappies and tea bags), with numbers predicted to rise by another 40 per cent by 2025.

The repeated ingestion of waste microplastics by marine life is even beginning to cause problems beyond the clogging of stomachs, as some toxic compounds found in plastic products have been detected as concentrates in fish tissues. Ironically, to close the loop on the cause of plastic pollution (ie humans!), such toxic plastics have been found in European anchovies, mackerel, striped bass, canned sardines, Pacific oysters and many more species that are sold for human consumption. How long before these toxic compounds start to accumulate in human tissue, too? Our food chain can therefore be said to be at risk from the very packaging in which we have chosen to serve it for too long.

Harnessing the social network

While all this can be seen as depressing, it is worth pausing to consider the potential value to ocean conservation of one of the more prevalent technological advances of the past decade. The Great Ocean Cleanup project's Twitter feed has nearly 100,000 followers. The feasibility phase involved a voluntary team of nearly 100 scientists and engineers. Over £1.6 million of the project's development budget was collected from the general public in an online crowdfunding campaign that reached its total in 100 days, eliciting donations from 38,000 funders in 160 countries. American broadcaster Oprah Winfrey encouraged the campaign online.

Protecting the ocean is receiving great interest in the age of social media. While earnest proclamations such as 'save the whale' have been high on the social agenda since public campaigns started to draw wider profile in the television age, the continued rise of new media craves head-turning imagery to attract attention, and ocean conservation has plenty such collateral. Each of the stories about finding dead whales or turtles ostensibly killed by ocean plastic pollution are accompanied by vivid images of the amount of waste found in their bellies, or of pictures of beaks lethally constricted by varying styles of plastic packaging. The most ground-breaking of all of the images that helped the ocean plastic pollution issue break through so compellingly into the public consciousness was a moving one.

The final episode of the BBC's *Blue Planet II* series, presented by Sir David Attenborough and broadcast in December 2017, catapulted public awareness of ocean plastic pollution to unprecedented levels. A scientist from the British Antarctic Survey stood over the open carcass of an albatross, revealing that a plastic tooth-pick had stuck in its stomach and killed it. A turtle was shown caught helplessly in discarded plastic fish netting. The scene that evoked the most visceral response, though, was the sight of a distressed female pilot whale, which was shown to be carrying something white in its mouth. Closer inspection revealed it to be her dead calf. The mother was filmed carrying it for several days, seemingly unable to let go. The cause of death was attributed – potentially, not definitively, it should be noted – to poisoning as a result of an increase in microplastics in the pod's food chain, passed through to a juvenile that could not counter its toxicity at such a fragile stage in its life.

The ensuing public outpouring on social media sites (often accompanied by the arresting shareable video content or imagery) were typified by messages such as the following on Twitter: 'Never using a plastic bag ever again'; 'I'm nearly crying... my God we must do something about plastics in our oceans'; and 'I hope every person watching *Blue Planet* makes even the smallest changes to try and reduce their plastic usage.'

Such enhanced public interest in this specific ocean conservation issue provides greater leverage to those trying to make a difference. While initiatives such the Great Ocean Cleanup lead the way at sea, the focus of many of the solutions (technology-driven or otherwise) for this ocean problem actually remains on dry land. As a population, we have understood for many decades that we should 'reduce, re-use and recycle', but greater knowledge of the burgeoning impact of the continued rise of plastic on the oceans appears to be acting as a catalyst. The emerging circular economy movement aims to move from a system whereby consumer goods have a shelf life and are then discarded to a fundamental rethink of how companies can deliver 'cradle-to-cradle' products that can be used again and again in the manufacturing cycle, rather than most of the components (some, campaigners would argue, actively designed to stop working within a certain period of time to drive new purchases) ending up in landfill. This movement is led ably by the round-the-world sailor Dame Ellen MacArthur, who in 2005 broke the world record for the fastest solo circumnavigation of the globe. Her understanding of the finite amount of resources available in the world was, she says, inspired by her sailing career.

Increasing public interest in ocean pollution has placed renewed pressure on producers of consumer goods to reduce the amount of unnecessary plastic packaging. Restaurants and bars are coming under increased scrutiny for relying on single-use plastics such as drinking straws. Plastic bag use has dropped significantly since governments introduced legislation that demanded that retailers had to start charging for them. Funders are responding to public and scientific interest in this field by creating new anti-plastic innovation funds, such as the UK's £20 million Plastic Research and Innovation Fund. The Ellen MacArthur Foundation and HRH The Prince of Wales' International Sustainability Unit also manage a £1.5 million New Plastics Economy Innovation Prize that challenges designers, entrepreneurs, academics and scientists to keep plastics 'in the economy and out of the ocean'.

On the policy and regulation front, Italy's Tremiti Islands have banned single-use plastic such as disposable cutlery, cups and plates, with fines of up to €500 for transgressors (Coffey, 2018). The Indonesian island of Bali introduced a similar ban in December 2018, which the local governor is hoping will lead to a 70 per cent reduction in the amount of plastic waste found in local waters (The Straits Times, 2018).

In the same month, the European Union agreed to ban some single-use plastics in an effort to cut marine pollution (France 24, 2018). Once this is formally ratified, individual European Union states will have up to two years to legally prohibit items such as disposable cutlery, plates, straws, plastic cotton buds, drink stirrers and balloon sticks, and start to introduce measures to reduce single-use plastic and polystyrene food and beverage containers. Perhaps the African nation

Rwanda leads the way on the regulation front, though, as over a decade ago they made it illegal to import, produce, use or sell plastic bags and plastic packaging except for specific industries. So, if you're caught in Rwanda with a plastic bag, you are committing a crime and could be fined, jailed or forced to make a public confession.

Not all counter-measure efforts are restricted to land, though. In October 2018 the International Maritime Organization adopted a new action plan and supporting measures targeted at reducing marine plastic litter from ships (Imo.org, 2018). Specific identified measures in the plan include a new study to quantify the size of the problem of marine plastic litter from ships; exploring the potential to make the marking of fishing gear mandatory, so that any 'ghost gear' that is dumped and becomes a floating hazard can be traced back to the perpetrators.

Going with the flow

As with most Blue Economy domains, satellite imaging will have a significant role to play. It has the potential to identify the size of the problem in greater detail, and satellite communications and high-powered computer modelling are already being used to analyse and then predict patterns of movement of plastic as it emerges from rivers and enters the sea. In 2018, researchers at the University of Oldenburg in Germany fitted low-cost 7 x 5 centimetre SPOT Trace devices, which connect to the Globalstar low Earth orbit satellite constellation, to floating buoys in the North Sea (Ocean News & Technology, 2018). The tracking devices comprise an integrated GPS receiver, transponder and motion sensor, and following their progress allows researchers to accurately track drift patterns, aided by sophisticated modelling tools and 3D visualization. This in turn allows them to understand in greater detail how the movement of plastic debris on the surface of the sea is affected by variable wind, current and tides.

Early findings confirmed that such research can help to identify the source of plastic litter, and to predict where best to target clean-up operations. This element looks especially useful as the early modelling provided scientists with data that countered the current thinking on how drift movement takes place in the study area, as wind at sea appears to have a greater effect on patterns than previously realized.

Autonomous ocean plastic collection devices are becoming plentiful. Like miniature versions of the much larger Great Ocean Cleanup system, examples include Captain Trash Wheel (given a comedic human appearance with the placing on it of googly eyes), a static device launched in Baltimore in 2018 that automatically collects waste from rivers and streams before it can flow into the harbour and thence the sea (O'Dowd, 2019). Taking up the baton from its prototype predecessors Mr Trash Wheel (in 2014) and Professor Trash Wheel (in 2016), the three wheels have collectively removed more than 680,000 kilograms of waste from the water.

The WasteShark is an autonomous marine litter collection system designed for ports and harbours. Developed by Dutch company RanMarine Technology and being trialled in the Port of Rotterdam, autonomous operation is mapped by waypoints, allowing it to suck up to 200 litres of waste out of the water in one trip, and to operate for up to 16 hours a day (KCI, 2018). It will also collect and send environmental data back to the cloud as it does so. Many similar systems are in development, and once fully tested they may become familiar sites in the world's major ports and harbours.

While considering such potential benefits, and reflecting that unmanned and autonomous systems are featured throughout this book, it's also important to point out that some researchers are keen to highlight that their broader effect on marine life is only just beginning to be understood. In 2017, for example, a research team at the Monterey Bay National Marine Sanctuary highlighted the potential drawbacks of the use of unmanned aerial vehicles at sea. Airborne drones are cheaper, more prevalent and more readily available to the average consumer than unmanned surface vessels capable of being deployed at sea. This means that many more will be used by hobbyists who may not be as informed as they should be on relevant civil aviation or environmental policy. Local to the Monterey researchers, that includes the need to adhere to the Marine Mammal Protection

Act, which prohibits disturbing marine mammals, be it by kayak, helicopter or – increasingly – aerial drones (Hoshaw, 2017). Research has shown that seals, sea lions and walruses are susceptible to noise disturbance from aerial drones operating at low altitudes. While those drones may be deployed in order to support useful oceanographic data collection, the concern is that – however well-intentioned – they could disturb feeding or rest patterns if not properly regulated.

Rather than banning them outright, though, it is likely that more education and public awareness campaigns will be launched in the coming months and years, such as the accessible tips to manage marine data collection or observation by drone offered online by the AliMoSphere project in California (Gaworecki, 2018). They have worked with local agencies to develop guidelines for the responsible deployment of unmanned aerial systems that fly close to wildlife near coastlines and the ocean. Such schemes will hopefully lead to many more drones (and, indeed, marine USVs, as and when they become more affordable) being deployed safely alongside more conventional marine science techniques. This is especially important as advanced technology becomes more affordable and moves out of sole commercial or academic use and into the hands of the average enthusiastic member of the public.

Power to (and from) the people

Technology thus creates new opportunities for the amateur oceanographer or – as they are increasingly becoming known – the citizen scientist. This interesting field has been growing in prominence in the past decade, aligned in the main to the exponential rise of smartphone ownership in all parts of the world. If citizens now carry in their pocket at all times what until very recently would have been considered a super computer, many innovators are developing ways for it be used to collect data, tackle micro-tasks, or other actions that can have a positive social impact.

That might just mean taking a photo of something that needs to be fixed in the local community and sending it through an app to the responsible authorities. However, other approaches are emerging that allow citizens to contribute directly to larger scientific investigations. One exciting early example in another domain came about when Cancer Research UK successfully harnessed mobile phones and the efforts of committed volunteers to help analyse millions of digitized human cells. This task would have taken them many months using traditional analysis techniques, but they achieved a much quicker result by turning the cell analysis into a space-based mobile shoot-'em-up game, which they called 'Genes in Space'. Players were instructed to 'zap' cells/'asteroids' that looked abnormal as their spaceship encountered them in space. This in turn meant that the cells that received the greatest number of 'hits' were most likely to be abnormal, so were accelerated for further analysis. This is where the average citizen attains greater significance as part of the wider 'wisdom of the crowd'. One might not expect a medical research organization to put their faith in the analytical skills of a member of the public with no previous experience in abnormal cell assessment. However, when 10,000 such individuals all identify the same cell or cells as being worthy of further analysis within the stated guidelines (when many others are not targeted), such analysis-at-scale begins to provide comfort.

Within the Blue Economy, harnessing the sensor and processing power of citizens' mobile phones – as well as their volunteer effort – has recently been applied in coastal southern Virginia, to allow interested members of the public to help improve the mapping of tides. The King Tide floods come every year to the Hampton Roads area, and local scientists are always keen to observe specific activity and flood characteristics in as much detail as possible. Starting in 2017 a smartphone app created by Virginia nonprofit Wetlands Watch increased observation capacity significantly as it allowed groups of volunteers to submit GPS coordinates and photos of floods to local project leaders (Dietrich, 2018). Over the allotted weekend, more than 500 people used the app to crowdsource data by taking and submitting pictures and GPS coordinates, often providing data that the scientists could not reach due to blocked transport routes. Local online, print and television partners helped to spread the news of the 'Catch the King' campaign, which ultimately led to more than 53,000 individual data points being collected in one morning. This deluge of new authoritative information enabled local authorities both to calibrate new water-level sensors that had just been installed and to improve the accuracy of their predictive modelling systems to assist with preparation for future flooding events. The website of the app's developers, Sea Rise Solutions, now lists hundreds of locations globally where local groups are putting this technology to use.

Another example is the Seagrass Spotter, an app that also allows ocean enthusiasts around the world to become citizen scientists by recording and sharing instances of seagrass via a tailored smartphone app. Promoted jointly by Cardiff and Swansea Universities and independent charity Project Seagrass, it aims to establish a much broader and more sustainable monitoring network than currently available solely within scientific resource by encouraging fishers, scuba divers and holidaymakers to use their smartphones to record locations of this threatened resource (Cardiff University, 2017). The site's gallery features contributions from places as diverse as County Kerr in Ireland, Laamu Atoll in the Maldives and the Marlborough region in New Zealand. If each of these images can be assumed not to have been available without this app, you can see how such mapping adds to global understanding of the prevalence and relative health of various types of seagrass, and perhaps suggest new research projects or inspire new practical conservation measures to help protect threatened ocean habitats. All made possible because of digitally driven citizen science.

Looping back to the ocean plastic pollution problem, the Global Ghost Gear Initiative (a collaborative effort to find solutions to the problem of lost, abandoned and otherwise discarded fishing gear – much of it plastic) has developed a Ghost Gear Reporter app for iOS and Android that allows seafarers to share the location and type of any gear they encounter (UN Environment, 2018). This potentially provides useful additional help on a problem the UN says leads to 640,000 metric tons of fishing gear going missing at sea each year.

Sounding off

Blue Economy citizen scientists are not just restricted to mobile data collection, however; the Orcasound project off North America's north-east Pacific coast allows them to be important subsea signal detectors to assist with whale conservation. The project has established a network of underwater microphones (known as hydrophones) that are placed at depths of approximately 26 and 98 feet in the Salish Sea off the coast of Washington State and British Columbia, Canada (OrcaSound, 2019). They then developed a web-based application (accessible by laptop, smartphone or tablet, the app for which was made possible through a $20,000 online Kickstarter crowdfunding project) that allows citizen scientists to listen out for the sounds of killer whales (Orcinus orca) on the live streaming audio feed. Listeners are tasked with reporting any unusual sounds.

As well as raising awareness and understanding about ocean noise, data provided also contributes to the task of decoding orca language. There's also evidence that sonar has caused mass strandings in some parts of the world, so users are instructed to listen out for significant noise pollution incidents so their source can be identified and mitigating actions taken before any harm comes to the local whale population. Users are 'trained' by listening to previously recorded calls of specific whale pods, so they can learn to understand what they are hearing. This is another example of humans and advanced technology working in tandem, as the efforts of the citizen scientists complement the development of machine-learning algorithms.

A major report published in the January 2018 issue of the journal *Ocean and Coastal Management* revealed that the general public widely believes that the marine environment is under serious threat from human activities, and supports actions to protect the marine environment in their region (ScienceDaily, 2018). Based on an impressive data set – over 32,000 people in 21 countries were surveyed – the results show that 70 per cent of respondents believe that the marine environment is under threat from human activities, with 45 per cent believing that the threat is high or very high. Unsurprisingly when considering what we have covered in this book, pollution and fishing were identified as the key threats, followed by alteration of habitats, climate change and loss of marine biodiversity.

It's interesting to note that the ranking of these threats by the public runs counter to what scientists generally consider to be the major challenges to ocean sustainability. While the study showed that the general public perceived pollution as the highest threat, scientific evidence has generally prioritized fishing and habitat loss as the highest threats to marine ecosystems. While there may be debate about the ranking of those threats, though, greater awareness of sustainability issues in the general public obviously provides greater support for lobbying activities and intervention strategies across the board. And this is further supported by celebrity endorsement. Globally famous people such as Leonardo Di Caprio (via his own charitable foundation), Harrison Ford and Barbra Streisand (named as celebrity supporters of US NGO Oceana), and Richard Branson (through Ocean Unite, associated with the Virgin brand) all publicly support ocean conservation, issuing rallying calls to their many millions of social media followers as well as leading particular funding and intervention campaigns.

Pushing the boundaries

Of course, more traditional scientific methods, ie not involving citizen scientists, are also abundant, building on centuries of tradition and rigour yet always pushing the boundaries of technology in an attempt to counter new challenges.

One of the major trends in the field of ocean conservation in recent years has been the establishment of Marine Protected Areas (MPAs). In November 2017, Mexico was the latest country to announce the establishment of a 'super MPA', the natural reserve protecting 57,000 square miles of marine life around the Revillagigedo Islands from fishing, mining and other extractive industries (Daley, 2017). The waters offshore are known to support 366 species of fish and the islands also function as calving grounds for humpback whales, and support coral gardens and other marine ecosystems.

This is just one of many such announcements, as the global ocean community successfully aligns to the headline UN Sustainable Development Goal 14 target of conserving at least 10 per cent of coastal and marine areas by 2020 (the Atlas of Marine Protection estimates that only 4.8 per cent of the ocean is currently protected in 'implemented and actively managed marine protected areas') (Mpatlas.org, 2019). Publicly welcomed by most in Blue Economy networks, the main challenge – for this and every MPA – remains enforcement. The more MPAs there are, the larger the area to cover. As our chapters on maritime surveillance (Chapter 6) and sustainable fisheries (Chapter 10) underline, a system of systems involving traditional human intervention, satellite-enabled services and autonomous or unmanned delivery of inspection may provide the best grounding for the daunting challenge of enforcement of MPA borders.

The US State of Hawaii and the Territory of American Samoa have started to move this way, as in October 2017 they announced a new technology partnership with the National Oceanic and Atmospheric Administration and Liquid Robotics, a Boeing company that provides unmanned surface vessels. Between them and other partners, they are deploying the Wave Glider USV for a range of environmental monitoring and surveillance activities, including focusing on illegal, unreported and unregulated fishing, water quality and marine debris, coral reef damage and climate change (Liquid Robotics, 2017).

The MPA prize is great. New research released in 2018 used hydroacoustic technology to find that the fish population was four times more abundant in Mexico's protected Cabo Pulmo National Park, which had been established in 1995, than the neighbouring regions. Researchers surveyed the marine life using sound waves produced by hydroacoustic equipment mounted on boats – a much more cost-effective method than underwater visual censuses using divers (Berger, 2018).

Enforcement aside, many other significant challenges still present themselves to the ocean conservation community below the water's surface. For example, no chapter on ocean conservation could possibly pass without mention of the growing threat of climate change to the Blue Economy. One of the main Blue Economy threats arising from the steadily increasing global temperature focuses on organisms already highlighted as being under threat from plastic within this chapter – coral reefs.

The continual absorption of carbon dioxide by the ocean is known to increase acidity levels. The UN Sustainable Development Goals 2018 update report stated that levels of marine acidity have increased by about 26 per cent on average since approximately the middle of the 18th century. When this is combined with the warming of the oceans, it can lead to a phenomenon known as coral bleaching, which has been witnessed in unprecedented levels in the past few years. When corals are stressed by significant changes in marine conditions, they can react by expelling the symbiotic algae that live in their tissues. This initially causes the corals to turn completely white and, if they cannot regain stability, eventually to die.

Starkly, some scientists predict that if the oceans continue to warm at the current rates, this temperature rise will eradicate coral reefs by 2050. If that seems to be scaremongering, it has been proven that the world has already lost roughly half its coral reefs in the last 30 years, so their total demise will just be the natural conclusion of the current trend (CBC, 2017).

The loss of reefs has an immediate knock-on effect on the ocean habitat of the species that rely on corals for food and protection. This creates further risk for coastal communities (at least 275 million people worldwide live near coral reefs), especially in low-income countries, who rely on highly productive reefs for food and commercial revenue from fishing, and from tourism.

Reports elsewhere in the past months bring little relief. One study showed that over-fishing removes the natural predators of coral-eating snails, thus allowing them to grow unchecked into a major menace (Caruana, 2018). Invasive rats have also been identified as a major threat to tropical island coral reefs. About 90 per cent of tropical islands have invasive rats on them and they were shown in a 2018 study to threaten coral reefs (the study focused on the Chagos Archipelago in the Indian Ocean) by eating the young of sea birds whose guano fertilizes the reef when it is washed out to sea (Shukla, 2018).

Thus, new approaches to reef protection and restoration are needed more than ever. Thankfully, the third International Year of the Reef in 2018 provided lots of encouraging case studies of how technology is being marshalled to help mitigate the effects of climate change and other emerging threats to coral reefs and restore them to health (IYOR 2018, 2017).

Reap what you sow

One new tool showing promise of the required scalability is related to the process where marine scientists halt reef degradation by 'sowing' corals. Traditionally, this was an extremely time-consuming, expensive and laborious process (capturing coral seed during yearly mass spawning events; growing new corals in the lab and attaching by hand to existing, denuded coral). Ohio-based nonprofit Secore International is one company leading the way in developing this approach. They cultivate the coral larvae in specially designed star-shaped concrete 'seeding units' that allow them to simply be wedged into existing larvae, rather than having to spend much more time attaching them manually. The company estimates that this process – likened by some to farmers scattering seed on land – could reduce processing time by over 90 per cent, thus making it a more viable option for mass-scale adoption, which is undoubtedly what is required in the face of multiple threats (EurekAlert!, 2018).

Not surprisingly, artificial intelligence and autonomous systems have also entered the coral battle space. Scientists from the University of Queensland have been trialling a new AI system in Indonesia where 360-degree cameras fitted on underwater scooters allowed researchers to photograph up to 2 kilometres in a single dive (21st Century Tech Blog, 2018). Following a period of 'supervised learning', the recognition software starts to use algorithms and its own judgement to recognize corals and other formations of interest. The mass of collected images were subsequently analysed much faster than human scientists possibly could – reducing the analysis time from up to 15 minutes to a matter of seconds. This allowed swift evaluation of how global warming-coral bleaching between 2014 and 2017 had affected the area of study. As more coral bleaching occurs globally, such significant cost- and time-saving approaches can only be welcomed.

On the autonomy front, millions of coral-eating crown-of-thorns starfish (COTS) have a new adversary in the Great Barrier Reef – the RangerBot. A roboticist from Queensland University of Technology has developed this autonomous underwater vessel that not only has the ability to find COTS, but – thanks to recent, separate academic progress in understanding lethal threats to the starfish – can actually kill any COTS it encounters with a single injection of a derivative of bile (Braun, 2018). While only five RangerBots are currently in operation, this powerful mix of autonomy, recognition and robotics illustrates how such confluence of technologies bodes well for environmental conservation.

Australia led the way in encouraging humans to understand and counter the sun's harmful UV rays with their iconic 'slip-slop-slap' public awareness campaign in the 1980s, and they are now leading the way in preventative measures against the ocean warming incidents that can lead to coral bleaching. The Great Barrier Reef Foundation, University of Melbourne and Australian Institute of Marine Science have collaborated to develop an ultra-thin shield (50,000 times thinner than a human hair) to block the sun's rays (Phys.org, 2018). The biodegradable material takes the corals' own defences as inspiration, as it contains calcium carbonate, which is the same ingredient corals use to make their hard skeletons. It's designed to sit on the surface of the water above the corals, to provide an effective barrier against the sun. Initial testing was promising, as it reduced light by up to 30 per cent.

Conclusion

While we cover how technology is helping to encourage sustainable fisheries in Chapter 13, and touch on the emerging potential to identify oil slicks from space, ocean conservation is a much broader discipline, with many more challenges across the full breadth of highly complex marine ecosystems.

As we have seen, to take one example, while the world's corals face increasing levels of threat from a seemingly ever-expanding list of sources, they are receiving new levels and styles of protection from space, from autonomous systems, from artificial intelligence and from nanotechnology providing a very modern type of sunscreen.

However, perhaps the most powerful technology is that which underpins the global social media networks. They: enable mass messaging related to the problems facing the world's seas and oceans; inspire and drive thoughtful collaboration around what can be done to counter negative effects; encourage specific actions that can make a difference (be they digital or involve physical action in the 'real world'); and allow the celebration of success through shareable media that can recruit the next wave of volunteers and interested citizens.

The concept of crowdfunding is not exactly new. The call for the community to co-fund repairs to the church roof has been an ever-present part of village life for decades. Looking further back, the building of the granite pedestal of New York's iconic Statue of Liberty (itself a diplomatic gift to the US from France) was only made possible after a call for public donations was printed by publisher Joseph Pulitzer in his newspaper *The New York World*. The campaign raised the final $100,000 needed to complete the pedestal from more than 160,000 donors (with a little left over for a gift for the sculptor). More than three-quarters of the donations were for less than a dollar (BBC News, 2013).

Such appeals are now far more commonplace in an era when donors can not only receive news of appeals on their smartphone wherever they are, but also make their donations immediately thanks to embedded mobile payment options. As varying marine projects are well served both with beautiful imagery of pristine ecosystems that need to be protected and commensurately harrowing pictures of marine wildlife declining at the hands of human thoughtlessness, they have a steady stream of shareable content that can propel the ocean conservation call far and wide.

While the Great Ocean Cleanup has yet to prove itself fully, its young founder understood this power. By skilfully harnessing the story-telling reach of the TEDx platform, Dutch marine entrepreneur Boyan Slat took his story to many millions, and in doing so eventually raised $35 million and engaged the support of organizations as diverse as the Government of Netherlands, paint manufacturer AkzoNobel, dredging and heavy lifting company Boskalis, the University of Vienna, professional services giant Deloitte, and MAERSK, the world's largest container shipping company.

The smartphone revolution also potentially provides rocket fuel to the citizen science movement, where large numbers of the general public can be empowered to contribute to serious marine discovery or protection, often by digital means. The key to such initiatives – as with the launch of any mobile app – is discoverability, so we hope that larger organizations with broad networks and media reach, as well as celebrities with hundreds of millions of online followers, will collaborate with app innovators to launch and celebrate many more similar schemes.

At headline level, it was most notable when drawing together the content for this chapter that much of the innovative technology aligned to some of the more troublesome ocean conservation issues is being utilized to quantify the size and nature of the associated problems rather than actually to counter or mitigate against those problems. Yes, autonomous vessels are starting to collect plastic pollution and kill threats to coral reefs, and smartphones are empowering the general public to contribute to the ocean conservation cause in new ways. However, many of these initiatives are at quite an early stage and involve small-scale experiments, in contrast to the wider use of satellites and AI to deepen understanding of changes in the seas and oceans. In many ways we are only just coming to terms with the damage we are collectively unleashing on marine ecosystems globally, and scientists are utilizing new technologies in creative ways to keep up with a story of environmental damage that is perpetually changing.

If many of the challenges facing the ocean are man-made, then the solutions will have to be, too. It is therefore hoped that the striking visual nature of ocean conservation issues, which provides perfect shareable content across ever-expanding social media networks, can corral many more to the conservation cause, to increase the chance of success and reverse the decline we see in so many parts of the world. Time may be running out for interventions at scale.

References

21st Century Tech Blog (2018) 360-degree underwater cameras and artificial intelligence may save coral reefs. www.21stcentech.com/images-artificial-intelligence-save-coral-reefs/ (archived at https://perma.cc/P3CR-NJZZ)

Applied Sciences from Technology Networks (2017) Taste, not appearance, drives corals to eat plastics. www.technologynetworks.com/applied-sciences/news/taste-not-appearance-drives-corals-to-eat-plastics-293620 (archived at https://perma.cc/9GR3-ZGNK)

Bas.ac.uk (2018) Increase in plastics reaching remote South Atlantic Islands. www.bas.ac.uk/media-post/increase-in-plastics-waste-reaching-remote-south-atlantic-islands/ (archived at https://perma.cc/A9CJ-5UXG)

BBC News (2013) The Statue of Liberty and early crowdfunding. www.bbc.com/news/magazine-21932675 (archived at https://perma.cc/RD48-YW8B)

Berger, M (2018) How a scientist counted the fish in a huge marine reserve in just 8 days. www.newsdeeply.com/oceans/articles/2018/01/30/how-a-scientist-counted-the-fish-in-a-huge-marine-reserve-in-just-8-days (archived at https://perma.cc/G8PR-7YQQ)

Braun, A (2018) RangerBot: Programmed to kill. www.hakaimagazine.com/news/rangerbot-programmed-to-kill/?xid=PS_smithsonian (archived at https://perma.cc/4GWB-VECF)

Burgess, A (2018) Green Sea Guard: Supporting energy transition in the shipping industry. www.governmenteuropa.eu/green-sea-guard/90480/ (archived at https://perma.cc/22J3-X4YG)

Cardiff University (2017) Can citizen scientists locate the world's seagrass? www.cardiff.ac.uk/news/view/1014747-can-citizen-scientists-locate-the-worlds-seagrass (archived at https://perma.cc/EAV5-ASE8)

Caruana, C (2018) Overfishing fosters growth of coral-eating snails. www.scidev.net/asia-pacific/fisheries/news/overfishing-fosters-growth-of-coral-eating-snails.html (archived at https://perma.cc/7XFC-H2FV)

CBC (2017) Scientists race to save world's coral reefs. www.cbc.ca/news/technology/scientists-race-prevent-coral-reef-wipeout-1.4022248 (archived at https://perma.cc/XAJ5-TSDK)

Coffey, H (2018) Visitors bringing plastic utensils to these Italian islands will be fined up to €500. www.independent.co.uk/travel/news-and-advice/italy-islands-plastic-ban-isole-tremiti-archipelago-fines-ocean-pollution-tourists-a8329631.html (archived at https://perma.cc/KD9D-CF9E)

Daley, J (2017) Mexico establishes largest marine protected area in North America. www.smithsonianmag.com/smart-news/mexico-declares-north-americas-largest-marine-reserve-180967309/ (archived at https://perma.cc/U6MG-FWGB)

Dietrich, T (2018) Hampton Roads is invited to 'Catch the King'... again. www.dailypress.com/news/science/dp-nws-evg-catch-the-king-sequel-20180822-story.html (archived at https://perma.cc/2QD4-MSEE)

Ellis-Petersen, H (2019) Dead whale washed up in Philippines had 40kg of plastic bags in its stomach. www.theguardian.com/environment/2019/mar/18/dead-whale-washed-up-in-philippines-had-40kg-of-plastic-bags-in-its-stomach (archived at https://perma.cc/VTF6-VFM6)

EurekAlert! (2018) Sowing corals: A new approach paves the way for large-scale coral reef restoration. www.eurekalert.org/pub_releases/2018-01/si-sca010318.php (archived at https://perma.cc/HCA9-YBD2)

France 24 (2018) EU officials agree to ban single-use plastics. www.france24.com/en/20181220-eu-officials-agree-ban-single-use-plastics-sea-pollution (archived at https://perma.cc/2Q5A-PAH7)

Gabbatiss, J (2018) Plastic is 'killing corals' by increasing risk of disease in reefs, say scientists. www.independent.co.uk/environment/plastic-pollution-coral-reefs-disease-damage-seas-oceans-cornell-university-a8178156.html (archived at https://perma.cc/U9SY-TE54)

Gaworecki, M (2018) Audio: How to use drones without stressing wildlife. https://news.mongabay.com/2018/07/audio-how-to-use-drones-without-stressing-wildlife/ (archived at https://perma.cc/HWY7-XMX7)

Hoshaw, L (2017) Drones on Monterey beaches frighten seals during mating season. www.kqed.org/science/1457720/drones-on-monterey-beaches-frighten-seals-during-mating-season (archived at https://perma.cc/6QKJ-TKFF)

Imo.org (2018) Addressing marine plastic litter from ships: Action plan adopted. www.imo.org/en/MediaCentre/PressBriefings/Pages/20-marinelitteractionmecp73.aspx (archived at https://perma.cc/54RD-XJCR)

IYOR 2018 (2017) 2018 is International Year of the Reef. www.iyor2018.org/news/2018-is-international-year-of-the-reef/ (archived at https://perma.cc/4V2T-QALA)

Kao, E (2018) 17 million pieces of plastic a year flushed into sea via Shing Mun River. www.scmp.com/news/hong-kong/health-environment/article/2169129/more-17-million-pieces-plastic-waste-flushed-sea (archived at https://perma.cc/5FN4-GGJF)

KCI (2018) Chomping down on trash: WasteShark collects marine debris. www.kci.com/resources-insights/bluecurrent/chomping-down-on-trash-wasteshark-collects-marine-debris/ (archived at https://perma.cc/3X82-459N)

Liquid Robotics (2017) NOAA's Office of National Marine Sanctuaries and Liquid Robotics collaborate to protect vulnerable marine sanctuaries and ecosystems. www.liquid-robotics.com/press-releases/noaas-office-national-marine-sanctuaries-liquid-robotics-collaborate-protect-vulnerable-marine-sanctuaries-ecosystems/ (archived at https://perma.cc/6FHH-WPZR)

Marine Insight (2019) The ocean cleanup system 'Wilson' to return to port for repair and upgrade. www.marineinsight.com/shipping-news/the-ocean-cleanup-system-wilson-to-return-to-port-for-repair-and-upgrade/ (archived at https://perma.cc/VY8H-KMQL)

Mpatlas.org (2019) Home page. www.mpatlas.org/ (archived at https://perma.cc/H48U-JXPD)

Neves, D, Sobral, P. Lia Ferreira, J and Pereira, T (2015) Ingestion of microplastics by commercial fish off the Portuguese coast. *Marine Pollution Bulletin*, 101. www.researchgate.net/publication/284275589_Ingestion_of_microplastics_by_commercial_fish_off_the_Portuguese_coast (archived at https://perma.cc/Q348-ZBJX)

New Straits Times (2018) Whale washed up in Indonesia had 6kg of plastic waste in stomach. www.nst.com.my/world/2018/11/433059/whale-washed-indonesia-had-6kg-plastic-waste-stomach (archived at https://perma.cc/UXH3-9X4B)

Ocean News & Technology (2018) Globalstar Satellite technology to tackle North Sea plastic pollution. www.oceannews.com/news/communication/globalstar-satellite-technology-to-tackle-north-sea-plastic-pollution (archived at https://perma.cc/RN25-ZQRN)

O'Dowd, P (2019) Meet Mr Trash Wheel: Baltimore Harbor's googly eyed garbage gobbler. www.wbur.org/hereandnow/2019/04/16/mr-trash-wheel-baltimore (archived at https://perma.cc/8ME6-5VR5)

OrcaSound (2019) Orcasound app. www.orcasound.net/portfolio/orcasound-app/ (archived at https://perma.cc/5PZD-3MDD)

Phys.org (2018) Ultra-thin sun shield could protect Great Barrier Reef. https://phys.org/news/2018-03-ultra-thin-sun-shield-great-barrier.html (archived at https://perma.cc/WD78-VQS6)

ScienceDaily (2018) Strong support for ocean protection: Study. www.sciencedaily.com/releases/2018/01/180110101008.htm (archived at https://perma.cc/QQ8J-4ATX)

Shukla, P (2018) The newest threat to coral reefs: Rats. www.forbes.com/sites/priyashukla/2018/07/25/the-newest-threat-to-coral-reefs-rats/#14fa28a65004 (archived at https://perma.cc/7A97-XMMS)

The Maritime Executive (2017) Just ten rivers contribute most plastic pollution. www.maritime-executive.com/article/just-ten-rivers-contribute-most-plastic-pollution (archived at https://perma.cc/KYA4-D6ZE)

The Ocean Cleanup (2018) The world's first ocean cleanup system launched from San Francisco. www.theoceancleanup.com/press/the-worlds-first-ocean-cleanup-system-launched-from-san-francisco/ (archived at https://perma.cc/H7V3-GMXK)

The Straits Times (2018) Bali bans single-use plastics in bid to cut rubbish in sea. www.straitstimes.com/asia/se-asia/bali-bans-single-use-plastics-in-bid-to-cut-rubbish-in-sea (archived at https://perma.cc/CUS9-NAY8)

UGA Today (2015) Study: Stunning amount of plastic waste in the oceans. https://news.uga.edu/study-stunning-amount-of-plastic-waste-in-ocean/ (archived at https://perma.cc/V8RY-DXUJ)

UN Environment (2018) How to banish the ghosts of dead fishing gear from our seas. www.unenvironment.org/news-and-stories/story/how-banish-ghosts-dead-fishing-gear-our-seas (archived at https://perma.cc/PP6T-BV86)

Watts, J (2019) World's deepest waters becoming 'ultimate sink' for plastic waste. www.theguardian.com/environment/2019/feb/27/worlds-deepest-waters-ultimate-sink-plastic-waste (archived at https://perma.cc/7TWT-6AUX)

Wyatt, T (2018) Exotic Japanese fish hitchhiked 5,000 miles on tsunami to California, scientists say. www.msn.com/en-sg/news/world/exotic-japanese-fish-hitchhiked-5000-miles-on-tsunami-to-california-scientists-say/ar-BBQUQTW (archived at https://perma.cc/YZB4-LEJN)

Zachos, E (2018) How this whale got nearly 20 pounds of plastic in its stomach. https://news.nationalgeographic.com/2018/06/whale-dead-plastic-bags-thailand-animals/ (archived at https://perma.cc/8V7C-V84U)

10

Sustainable fisheries

Casting a digital net to help feed the world

The world relies on fish like never before. In the 1960s, the world's population consumed just under 10 kilograms of fish per person per year, a figure that had more than doubled to over 20 kilograms per capita by 2016 (Fao.org, 2018).

As explained in greater detail in Chapter 7, the continued rise of the aquaculture sector in the past few decades have provided much needed capacity here, but new challenges are emerging all the time. Not the least of those challenges, as identified by the Food and Agriculture Organization (FAO), is the prediction that combined production from capture fisheries and aquaculture will reach 201 million tonnes by 2030, an increase of some 18 per cent from 2016 levels (Fao.org, 2018).

That growth projection presents challenge enough to the world's fishing sector and associated environmental agencies, but it is rendered additionally problematic when current fish sustainability levels are taken into consideration. More than a third of the major commercial fish species that the FAO monitors is fished at biologically unsustainable levels, with 60 per cent being fished at biologically sustainable levels and the remaining 7 per cent under-fished. These concerning figures attract an additional layer of alarm when you consider that the global share of marine fish stocks considered within biologically sustainable levels stood at 90 per cent in 1974, a very worrying decline.

In order to ensure that activity over the next decade and a half does not push sustainability levels further into the danger zone in various parts of the world, the FAO highlights a number of areas that require urgent and persistent attention. Prominent amongst these is the demand for greater efforts to reduce the astonishing amounts of fish that are captured and subsequently discarded at sea.

Commercial fishing fleets need to provide what the market demands. This means that they are looking to capture, process and sell specific types of species, at the right level of maturity, and thus anything else that they pull out of the sea is often nothing more than an inconvenience to the operation of their business. This is known as bycatch and means that many types of fish and – critically – juvenile fish who have yet to mature to a state where they have neither produced offspring or are saleable when processed can be simply thrown overboard, or discarded. If this happens regularly, the loss of large numbers of juvenile fish threatens the ecosystem and puts stock sustainability at serious risk.

The FAO estimates that between 8 and 25 per cent of the total global fisheries catch is discarded each year, with the cost of those discarded fish considered to amount to around £1.38 billion per year. Added to this, of course, those needlessly killed juvenile fish do not go on to mature and strengthen species stock numbers through further reproduction. This issue is part of a much larger problem of inefficient and short-sighted fisheries management globally, with World Bank estimates suggesting that poor management and over-exploitation of fishery resources costs the global economy £38 billion every year. The FAO estimate that one in three fish caught around the world never makes it to the plate, as they are either thrown back overboard as discard, or rot before they can be eaten due to ineffective management practices. This widespread waste is especially important as roughly one-fifth of the global population relies on fish as its primary source of protein, so needlessly discarding 27 million tonnes of fish per year (as estimated by Oceana) presents a serious threat to a sustainable food source that is needed most by communities with limited other options.

Economic and social issues aside, there are a number of ecological problems associated with bycatch and discards, which have been shown to threaten damage to the broader biodiversity of the oceans. Directly related impacts recorded include dangerously reduced levels of oxygen (known as anoxia) of the seabed environment due to excess organic loading from discards; the disruptive increase of many scavenger species attracted by rotting fish carcasses routinely cast overboard; and threats to many non-fish species that are relevant to the functioning of the overall ecosystem (eg turtles, dolphins) that are unfortunately caught in huge commercial nets.

Within the complexities of the broader marine ecosystem, the combined mass of discards eaten by seabirds has been estimated to account for more than the total amount of live fish they consume. Thus, it is suggested that fish discards support substantial bird populations, which in turn further prey on fish, adding another man-made circuitous route of threat to the sustainability of certain fish species.

The bycatch problem does of course vary in each fishery around the world. To take one example, salmon has been designated as a prohibited species in the Alaska commercial groundfish fisheries, and regulations require that bycatch of specific salmon species may not be sold. However, while commercial fishers can therefore attribute zero commercial value to Chinook, chum, coho, sockeye and pink salmon, they may still catch them in their huge nets by accident, in amongst the fish that that they are legitimately intending to catch. In that part of the world, bycatch is understood to be a particular problem in the groundfish fisheries in the Bering Sea and the Gulf of Alaska, especially on vessels fishing with trawl gear, but the nature, focus and extent of the issue globally is very much localized depending on local natural stocks and the focus of the activity of (and the gear used by) local fisheries operators.

In recent years the European Union, grasping the enormity of this problem, has increased the governance pressure on fishermen, owner-operators and their supply chains to identify ways to address these issues, with the introduction of the various discard plans for selected fish species. This has been matched by various fisheries management initiatives related to landing obligations and measuring and monitoring of Total Allowable Catch (quotas). However, in attempting to grapple with these growing regulatory demands, without necessarily challenging the need for them, some in the fishing industry have complained that credible, proven technical solutions and products to drive rapid system innovation in fish capture processes do not exist.

A light in the blue

Serious bycatch and discards innovation is therefore in demand, and UK start-up SafetyNet Technologies is attempting to commercialize the very latest science to address this urgent need. Based in London, this company manufactures interactive underwater lighting systems that have been designed to easily fit on to different types of fishing gear in wide use in the industry.

It is not widely known, but science has proven that varying patterns and colours of light can attract and repel different species of fish. This is a significant boon to fishermen who very much want to be increasingly selective about the fish they encounter during the capture process. Attempting to improve the sustainability of fisheries by offering the holy trinity of significant commercial, social and environmental return on investment all at the same time, the system modifies the colour and intensity of the light it emits in order to target the 'right' sort of fish. Amongst several products available, one of the most promising fashions these lights into rings that are built into trawler nets. The rings are designed to be the right size to accommodate the sorts of juvenile fish the trawlers don't wish to catch. Thus, when the lights pulse, the younger fish are attracted to it, and swim through the rings to the relative safety of the open sea, as opposed to being caught, killed and thrown back into the sea. More mature fish simply ignore the lights and stay within the catch.

In late 2016, SafetyNet worked in partnership with Young's Seafood to deliver early scientific testing that provided great encouragement. Young's is a major British producer and distributor of frozen, fresh and chilled seafood that supplies approximately 40 per cent of all the fish eaten in the UK, so were very keen to be involved in the research, which it made possible by providing selected trawlers. The research was led independently by the Centre for Environment,

Fisheries and Aquaculture Science, the UK Government's official fishery research arm. The trial took place in Farne Deeps North Sea fishing grounds and tested SafetyNet's innovative light rings, deployment of which was recorded to lead to an eye-catching headline reduction of 57 per cent in the number of small fish (at lengths under 24 centimetres) caught in the experimental trawl with the light-rings when compared with the standard trawl that was taking place at the same time in the same waters. There were noticeable variants across the differing species, but all figures were encouraging, with a 47 per cent reduction in the number of smaller Nephrops caught, 69 per cent of whiting, 39 per cent of plaice and a 58 per cent reduction in dab caught in the smaller length range.

Extremely encouraging signs, then, for the potential of an innovative technology based on scientific evidence and promising both environmental and commercial benefits, as reducing by 57 per cent the amount of juvenile fish caught in trawler nets means both that such fish do not have to be sorted out manually for disposal, and that the available net capacity can be filled with the desired fish, thus boosting efficiency and profitability. However, even though their product has been shown to have extremely promising environmental and commercial benefits, SafetyNet Technologies face the challenge of how to drive towards scale and break open a large and complex industry with many competing demands.

Providing the framework

As we see in many industries, and as referenced throughout this book, some customers view new technology options as over-hyped, pushed into the market before they are ready and sometimes not particularly well explained. New technologies can therefore be seen to be trading on the values, success and reputation of their predecessors or competitors, not just their own.

Within this context, emerging technology in the fisheries market is perhaps not helped by the fact there is so much of it. While it is obviously encouraging that the sector is truly bursting with new technologies being progressed to aid various elements of fisheries management, this can make it more difficult to stand out from the crowd and attract attention from a dazzled potential customer base.

One attempt to at least provide a broad framework within which each of these innovative approaches can thrive is provided by the United Nations, and particularly Sustainable Development Goal 14, which aims to 'conserve and sustainably use the oceans, seas and marine resources for sustainable development'. Ten targets and 10 indicators provide the headline focal points for related SDG activity. The global indicator framework was developed by the Inter-Agency and Expert Group on SDG Indicators (IAEG-SDGs) and agreed as practical starting points in March 2016 at the 47th session of the UN Statistical Commission. The framework provides a useful backdrop to the further strategic considerations of industry, funders, NGOs, start-ups, related UN agencies themselves and, increasingly, customer-facing corporations that wish to publicly celebrate their sustainable credentials in a competitive and increasingly transparent market place.

It is interesting to note that, while the targets cover a range of issues (from reducing marine pollution and ocean acidification to the establishment of marine protected areas), such is the scale of the fisheries sector that its activity can be said to play a pivotal role, or be significantly affected by, all of them. That ranges from the problem of lost or discarded fishing paraphernalia (known as ghost gear) affecting the marine pollution target (14.1); ocean acidification (14.3) having an adverse effect on fish stocks; and the target to accelerate the transfer of marine technology (14A).

Scanning across the fisheries sector, there is indeed much technology to transfer.

Within the trading area, companies such as Norway's JET Seafood are developing new tools to digitalize the fish trading process. The new platform, which focuses in the first instance on Norwegian salmon, is designed to enable sellers to log in and view what produce fishermen have available, then offer to buy stock at the agreed price or place a 'bid' if they think the advertised price is too high. Like commodities trading platforms in many other sectors, JET – whose founders are transferring the idea from the energy sector – will take a small percentage of each sale. The company says that this has the potential to replace the more traditional e-trading options of email, Skype or phone. In time, of course, as with any digital system, analysis of the data that flows through the system will no doubt start to provide new levels of value.

Many start-ups are applying their technical and industry know-how to develop digital catch documentation schemes. In December 2018, for example, the Government of Papua New Guinea announced its intention to strengthen its fisheries management capabilities by investing in a new system devised by its own national fishing authority. The Integrated Fisheries Information Management System aims to improve the ease with which authorities can collect detailed information on tuna caught in its waters, and related fishing vessel activity. It launched across a network of 240 purse seine (large netting) vessels, channelling real-time catch data from them to shore. Previously, it could take months for vessel owners to complete and deliver paper-based catch logs to the appropriate fisheries managers. Now, fisheries observers can also feed catch data into the system via electronic tablets.

As well as providing valuable data to allow sensible and timely management decisions about fish stocks and activity, the greater prize with such systems comes in their making possible much swifter and more accurate data on the provenance of fish. This is needed to attempt to tackle one of the fishing sector's greatest and longest-standing problems – that of illegal, unreported and unregulated fishing (IUUF).

A multi-faceted problem

Another of the key elements of the UN's SDG 14, Action 14.4 articulates the goal that, by 2020 and among other actions, collaborative global efforts will lead to the effective end of IUUF, specifically ending destructive fishing practices.

Variously estimated to be worth between £7.5 and £17.7 billion per year, the booming illegal fishing trade also denies (often poor) coastal communities the full value of their natural resource. Illegal fishers also care much less about the potential negative effects of bycatch, and can cause far greater damage to ecosystems by deploying destructive fishing practices, such as putting cyanide poison in the water or using the very blunt (but still productive) tool of dynamite to kill fish.

As with the potential impact of illegal fuel smuggling that was explored in Chapter 6, illegal fishing is a problem that has a broad impact. While the UN targets are articulated in SDG 14, the case can quite easily be made that negative impacts associated with IUUF can have significant adverse effects on at least 13 of the 17 SDGs. For example, securing fish stocks for local coastal communities will help to deliver SDG 2 (ending hunger); enhanced trading of fish will affect SDG 1 (ending poverty); protecting fishing activity across a nation will support sustained, inclusive and sustainable economic growth and full, productive employment (SDG 8). Tackling IUUF could also have a positive effect on the fight to counter climate change (SDG 13), as recent research has highlighted that large-scale ecosystem effects can occur as a result of predator removal, including increased production of biological carbon dioxide in the ocean (Stafford, 2016).

Improved digital catch documentation schemes can begin to tighten the net on illegal fishing practices as they promise to provide transparency right across the delivery chain, from the moment a fish is caught to the time it ends up on a plate in front of a consumer. With consumers becoming more ethically aware, major corporate providers are taking much greater interest in the promise of fish traceability schemes as they emerge.

One of the largest is Thai Union, based in the Thai coastal province of Samut Sakhon and boasting nearly 50,000 employees. In early 2018, the company announced the results of a small-scale traceability study that was conducted in partnership with the United States Agency for International Development (USAID) and Mars Petcare, one of Thai Union's major customers. Approximately 50 Thai fishers, operating on four separate vessels, were provided with electronic tablets and trained how to log their catch digitally. Data from this e-log was relayed back to shore in real time by specially installed satellite transmitters. The aim of the system is to provide Thai Union with the wherewithal to verify which fish are being caught where, and that they are being caught legally – in appropriately managed waters, on vessels with the right permits, etc. The constant connectivity that makes data transfer possible also allows crew members to connect with friends and family back on land, thus reducing the chances of the kind of human rights abuses featured in Chapter 12, by making ships less isolated while out at sea.

A wide range of such fish traceability schemes are now coming to market, all made possible by advances in mobile/tablet input technology and satellite connectivity options.

Vancouver-based ThisFish is another example of a working system, promising to accompany any fish in the supply chain with the details of 'who caught it, when, where and how'. This, they say, provides three immediate levels of benefit: cutting costs by replacing paperwork with digitalized data input; improving compliance by automating alerts on potential errors with real-time analytics; and increasing productivity by providing actionable data insights and allowing full transparent traceability.

A similar pilot programme was launched in 2018 by collaborators from the Fiji and New Zealand governments, WWF Australia, multinational blockchain specialist company ConsenSys, Fijian start-up/system provider TraSeable and tuna fishing and processing company Sea Quest Fiji Ltd. Many such systems have blockchain technology at their core.

Most often described as a secure digital ledger, blockchain originated in financial trading systems, but is now breaking out into most industries where the transparency of decentralized and verifiable data streams can provide value. Within the fish supply chain, this provides extra comfort as all agreed partners can access the data flows, but no individual can alter or delete the history of transactions to, say, cover up unauthorized or illegal activity. To take it to its logical conclusion in the fisheries sector, blockchain allows a consumer in a supermarket to scan a barcode and see exactly when, where, how and by whom a fish was caught. For the retailer, and everyone in the wider supply chain, this provides a very useful ethical stamp of approval at the point of purchase.

We expect to see a further proliferation of such blockchain-enabled technologies in the next few years, but some wider system issues will also need to be addressed in order for such systems to have proper utility across the market, not least – as with many Blue Economy issues – the large scale of business operations. In relation to fisheries, some commentators have highlighted that for blockchain-enabled traceability schemes to function at scale, it will require the trust, commitment, collaboration and potentially investment from a wide range of companies, governments and other stakeholders in many countries. That chain may involve fishermen, local markets, aggregators, primary processors, wholesalers, distributors and, finally, retailers. Each of these, potentially in multiple countries, would need to buy into the system to ensure its validity. While such mass buy-in will take time, and the developers of underpinning technologies will need to think and work internationally, the idea is attracting encouraging noises, especially from the consumer-facing companies who may have the best leverage to drive change through the system.

Airborne AI

As with most Blue Economy sectors, autonomous systems are being explored to assess their potential contribution to countering illegal fishing. Whereas we have previously explored how unmanned surface vessels are growing in significance in the maritime surveillance domain, airborne options are also coming to the fore.

Moroccan start-up Atlan Space is exploring how aerial drones can allow regulatory and enforcement bodies to scan thousands of square miles of ocean each day, specifically by aligning their aerial abilities with the targeting potential to be provided by artificial intelligence. Aerial drones aren't particularly new, but just sending a drone out to surveil waters could still require human piloting and guidance, thus – as we have discussed elsewhere – potentially negating the cost efficiencies that autonomous systems promise. By introducing artificial intelligence, though, the game begins to change.

The Atlan Space drone flies on a path determined by the automatic computer guidance system until the deep learning model identifies what it believes to be a boat. The system then analyses the image to identify whether it's a fishing vessel, then (using data such as the boat's name, flag and type of radio signals) attempts to determine whether it is legally permitted to operate in the region. The ultimate aim is to alert the relevant authorities to potentially unauthorized boats via a satellite message. More data processing is undertaken once the drone is back on dry land, both to strengthen potential vessel identification and to modify the computer modelling. The company suggests that a single drone can allow them to monitor 10,000 square kilometres a day, which would provide another useful tech-enabled tool for fisheries enforcement teams.

Ten thousand square kilometres is certainly an impressive coverage, but is still literally a drop in the ocean when you consider the vast expanses that need to be covered to identify and counter illegal fishing practices at sea, so again the sector is looking to complement near-surface activity with opportunities driven by satellites. The further progression of the on-orbit Earth observation and maritime surveillance infrastructure examined in Chapter 6 is providing great promise here. In its guide to Earth observation, the European Space Agency (ESA) underlines that rapidly expanding constellations of small satellites producing ever-increasing amounts of imagery will be highly likely to reduce costs for the resulting data. It also suggests that this eventuality may lead to a shift in the focus of related business models, as companies progress from simple provision of data to – as we have seen many times in examples in this book – more intelligent services further enabled by machine learning and AI. The land-based example ESA provides is a shift from the provision of simple images of shopping mall car parks to the potential for daily intelligence reports on the comparative popularity of shopping malls, by counting the number of vehicles in the car parks of each.

We have also spoken to other Earth observation companies who have received interest from insurers keen to explore its data to help them monitor construction projects. Why would investors wait for a quarterly report on the progression of a railway project cutting through the Indian subcontinent, when they can observe progress themselves using data from space? Similar ways of testing contractual compliance are also being discussed in agriculture, where it can be expensive to send observers regularly to rural or remote areas.

Within the fisheries domain, though, the intelligent use of Earth observation data is really beginning to gain traction. As briefly referenced in Chapter 6 on maritime surveillance, OceanMind is a UK-based fisheries compliance system that is designed to help fisheries analysts and maritime law enforcement agencies to adopt and maintain improved fisheries management practices. The company combines satellite data, powerful machine learning algorithms and accomplished fisheries enforcement experts to enable more effective countering of illegal, unreported and unregulated fishing by supporting governments and environmental agencies.

It can be tempting to view OceanMind as a simple vessel-tracking system, but that misconstrues the power and versatility of the underpinning technology. Many established organizations have the capability to track vessels and to analyse behaviours, but having a vessel tracking system is perhaps of limited use unless you know – and know how to apply – the relevant regulations in the specific seaspace in which the vessel is operating, and then are able to identify vessel activity or behaviours that may transgress those rules. Such rules and regulations change all the time – for example, as fishing seasons change, or to comply with new international policy. OceanMind provides a flexible system that combines as much data, from as many relevant sources as possible, to create knowledge that allows fisheries enforcement officials to do their job more effectively and more efficiently.

For example, the OceanMind system will first of all take raw or pre-analysed fisheries and vessel data in a specific location, which can then be cross-referred with the millions of relevant records it already has in its own databases – this could be anything from knowledge of local regulations to vessel-specific information such as previous convictions or even journeys marked previously as suspicious. Computing power then enables the 'crunching' of that combined data using hugely powerful machine learning algorithms to automatically identify vessel activity or behaviour that may warrant closer inspection. Finally, those alerts are fully interrogated by experienced fisheries compliance experts, who pass any live issues on to identified in-situ local partners for immediate action, or the development of longer-term policies and operations.

Each of these five stages (ingesting new data; combining it with existing knowledge databases; running algorithms to identify potential incidents of interest; interrogation by fisheries experts; and passing on to local actors) is important and warrants further exploration.

One of the OceanMind system's key strengths is that it is completely data agnostic. It is not a closed system reliant on the outputs of specific hardware, which can be the result of commercial companies wishing to make customers rely on its embedded products and data and analytical services. Instead it brings together the widest possible variety of data sources – including environmental, identity and positional sources – to understand the location, time and behaviour of specific vessels at sea.

As we see so often, new levels of intelligent computing power sit at the heart of possibility. The ability to ingest so many different sources of data provides a high-resolution, enduring database that alerts OceanMind fisheries experts to anomalies worthy of further investigation. This goes far beyond simple vessel activity. For example, port identification and potential transhipment hot spots have been automatically incorporated into the OceanMind solution. When satellite overpasses or other data inputs align at the right time, analysts are able to utilize a number of data sources to monitor vessels whenever they leave and return to port and can cross-reference vessel identity, such as flags and vessel numbers, and characteristics such as hold capacities to detect abnormalities. OceanMind also has the capability to identify unofficial ports, such as old landing piers and sheltered areas where transhipment activities may take place, to gain a full insight into regional fishing activities.

Some local data sources would have limited value on their own to the local enforcement agencies, but become powerful when absorbed into a system that drives complex correlation processes that produce actionable intelligence to help with local and international enforcement. Within the OceanMind system, this includes a wide range of potentially useful vessel activity information. Utilized data includes type of fishing activity (eg longlining, trawling, purse seining) and general vessel positioning activity (port visits, stationary at sea (more than 20 kilometres from shore); stationary near land (less than 20 kilometre from shore); stationary near port). Moreover, the system will provide alerts when it identifies missing data, for example when it identifies unrecorded movement between ports. Of perhaps even greater interest to authorities, proximity alerts are raised when vessels broadcast positions close together, which can indicate possible transshipments and bunkering (legal or illegal), which can be further corroborated by Synthetic Aperture Radar imagery.

When you combine all of these possibilities, what you get is a machine learning system that analyses not just vessel activity (simply what and where they are), but also vessel *behaviour*, intelligently combining data sources to provide clues as to what they may be doing, and to highlight anomalies that may be worth investigating. As an example, the OceanMind system is able to track vessel speeds, the patterns of variation of which can suggest specific types of activity. So, an alert may be sent when a fishing vessel slows to a speed that seems not in keeping with its stated activities. If a vessel that is supposed to simply be traversing a marine protected area is shown to be following a course that far more closely resembles fishing activity, that is worthy of investigation. To take that one step further, if a vessel turns off its automatic identification system just as it is entering a protected area, then reappears at a certain point later, the system can assess what speed it was doing during that 'dark' period, and align it to specific types of fishing activity.

OceanMind alerts are adjusted based on the needs of a given zone or fishery and can be paired with knowledge of the required local rules and authorizations where this information is available. All machine learning outputs are run against the location, time, place and permits to make sure the possible activity would be legal. All of this enables positional abnormalities to be identified and cross-checked against potential corruption of data, transmission gaps and abnormal vessel activity to detect possible illegal behaviour and flag it for analyst review.

Even as the Fourth Industrial Revolution becomes more pervasive, computers and machine learning can still only go so far. In OceanMind, experienced and highly skilled human experts are the final, vital interface between the huge computing power of the OceanMind system and positive local action.

OceanMind fisheries analysts – based in their office in Harwell in the UK, or on location in partner countries – have practical, 'in the field' knowledge and experience of fisheries monitoring and compliance. They have worked for coastguards, for Interpol, as fisheries observers actively boarding vessels for compliance observations and investigations. They understand the many hard complexities and intricacies of fisheries monitoring, compliance and enforcement, which is the final layer of intelligence that makes data actionable.

These analysts check to ensure that vessels are properly registered and licensed to fish in the area of interest and that these are both valid and in date. They look to see if vessels of interest have continuously transmitted on the required tracking systems and, specifically, they look for gaps in AIS transmissions. In addition to fishing in national waters, OceanMind analysts will also check to see if the vessel has fished on the high seas, or in the waters of another country and if so, attempt to verify that the vessel was licensed to do so.

They then look for layers of additional intelligence that can help to build out the picture. For example, a vessel seen to have turned into the wind and slowed to 1 knot for 10 hours could indicate behaviour that is long enough and slow enough to enable an at-sea transhipment. While no other vessels were detected on AIS, further satellite SAR data could be cross-referenced, and the imagery of the vessel itself checked for an indication that it can manage an at-sea transshipment. In some cases, suspicious activity might be explained away by weather events or an investigation into the commercial activity of the vessel. Any suspicious events that cannot be satisfactorily explained in this way, though, are detailed and any recommendations for further action by the relevant competent authority are given.

To recap, then, machine learning provides the ability to analyse quantities of data simply not possible if solely relying on human intervention, and produce automated alerts for further interrogation. OceanMind analysts then deepen the understanding of the incident using their own in-the-field knowledge and through further cross-referencing. This leaves local fisheries analysts doing what they should be doing – effective fisheries management and accurate, cost-effective tasking of monitoring, control and surveillance assets.

Because the intelligence system focuses on illegal activity, it is easy to imagine that much of the resulting actions will involve breathless, dangerous chases of potential criminals at sea, to catch them 'in the act'. While that can happen, more strategic options are available. First of all, it is far more cost effective to investigate the activity of individual vessels when they return to port or (in the case of international vessels) to the waters of their country of origin, where the Flag State can be contacted to share intelligence appropriately. Shore-side agencies will be provided with the collected intelligence to see if it matches up to what has been recorded in the ship's log.

More strategically still, taking a longer view of activity provides the potential to drive far more informed management practices. For example, by representing AIS tracks temporally and geographically, the system can provide 'heat maps' of activity within a country's waters. This can filter out commercial shipping and other types of sea-going vessel to just leave the focus on fishing boats. Mapping fishing activity (and, especially, where suspicious activity or potential transhipments are seen to occur) over several months can challenge or confirm extant understanding of key areas of operation, help to more effectively target sea patrols and elicit from the customer what additional monitoring or analysis they would like to see undertaken to provide next-level understanding of behaviours and potential threats. By undertaking such historical analysis, changes or trends in fishing activity within given waters can also be identified, and potential reasons explored.

More targeted reports can provide overviews of unlicensed fishing in areas of interest – where vessels are seen to be displaying fishing behaviours in the area of interest, but they do not have the correct licence (for example, one awarded by the Indian Ocean Tuna Commission) for all or some of the period. Again, such knowledge can aid patrol planning and allocation of other surveillance or the development of new intervention strategies.

The open data challenge

None of these systems are possible, of course, without the ever-expanding amounts of useful data that provide their intelligence engines with the fuel they need to make a difference. And that data can still be expensive.

We see that blockchain technologies are beginning to provide promise within the Blue Economy with their ability to provide the safest methods of transferring proprietary (closed) data. At the other end of the spectrum from this comprehensive locking away of data, though, there is another growing movement that aims to completely open it all up and in doing so significantly magnify its value.

Whereas previous debates have been dominated by considerations of who owns data (as well, of course, as the ubiquitous issues of data protection and privacy), the advent of the Fourth Industrial Revolution, driven as it is by the new potential of Big Data has opened up the potential to explore how governments could actually gain more by *giving away* data, rather than keeping it for themselves or attempting to sell it as a commercial offering. This is the open data movement.

In the UK, the Open Data Institute (co-founded by Sir Tim Berners-Lee, the inventor of the World Wide Web) has been working for the past few years to raise the profile of the commercial value and potential of open data, push the agenda for the standardization of data so that it can be more easily combined across data sets, and celebrate successful open data projects. Government agencies such as the Greater London Authority (the office of the Mayor of London) are investing in regularly refreshed open data hubs such as the London Data Store, to encourage commercial, social and academic data experts to make more of the information in over 700 data sets than they can themselves. Similarly, the World Bank has established a portal of open data to encourage those with an interest in global development to use new data analytics techniques to spot trends, identify anomalies and use data to inform new approaches to making a positive difference across the broadest range of social development disciplines. The United Nations supports a similar initiative in its UN Global Pulse programme. The leads of the programme see themselves as conveners – bringing data, analysts, governments and domain experts together to find the value that may make a difference to long-standing problems.

Commercial organizations are also beginning to explore how their proprietary data can be 'opened up' so that it can contribute to solving intractable social problems. Mobile phone companies are undertaking interesting work here. In recent years Orange, Telefonica, Vodafone, Telecom Italia and Digicell have all made their anonymized mobile phone data available for use in social development projects. Typically, this involves releasing 'synthesized data' that is based on a very detailed model of the network that simulates the activity of fictional customers whose behaviour is statistically like that of the real population. This therefore provides a detailed picture of mobile phone activity within a given area.

As has been shown in Chapter 5 with the likes of Westfield tracking footfall, mobile phone and Wi-Fi infrastructure can 'follow' the progress of a mobile phone (and, by extension, its owner) by recording where it connects to the network. While such individual data can often be useful in criminal cases to ascertain the movements of suspects or witnesses, it becomes even more powerful when aggregated. By combining millions of users' data records, you are in effect provided with a map of human traffic. When put in the hands of skilled data analysts with detailed knowledge of specific development issues (data is, after all, only useful if you know the right questions to ask of it), all manner of valuable insights can be gleaned.

In the Digicell experiment, researchers were able to understand the major routes the citizens of Port au Prince took when they were evacuating the city in the wake of the devastating 2010 earthquake. Brilliantly, by comparing the analysis with that of the previous

December, they were able to deduce that people went to the same places in the country (ie their extended family) as they had over the Christmas period. This is extremely helpful to plan future evacuation strategies.

Other projects have informed the planning of new public transport routes in Côte d'Ivoire cities (as analysts could see how far some citizens had to walk to reach available bus stops, so major gaps in provision could be identified) and the potential positioning of new fire stations in London, again to reflect human movement patterns and areas of greatest risk. In 2018, in a project originally inspired by needs identified as the country grappled with the 2014 West Africa Ebola outbreak, the Vodafone Foundation announced its intention to release millions of similarly synthesized mobile phone data records in Ghana, to help epidemiologists plan how any future outbreaks of disease may travel and best be managed. Whereas previously researchers would undertake occasional surveys at bus stops and similar places to try to understand the movement patterns of the population in order to plan containment strategies, they now have a rich data set of millions upon millions of movements across the country.

The same potential to derive real value from broad and varied data sets is obviously also present in the ocean environment, as underlined by the progress of OceanMind and others. For real progress to be achieved, however, the potentially prohibitive cost of such Earth observation data may need to be addressed.

Encouraging signs are undoubtedly beginning to emerge here. Global Fishing Watch is an open data platform (funded by the Leonardo Di Caprio Foundation, among others). One of their primary aims is to encourage governments and related agencies to make freely available data that can help the global fight against IUUF and support broader sustainable fisheries management. In 2017, Indonesia became the first nation to 'donate' data to Global Fishing Watch, when they made available the proprietary vessel monitoring system (VMS) tracking data from 5,000 smaller commercial fishing vessels that don't use AIS (as they are not required to). They were followed in October 2018 by Peru, who made available VMS data on 1,200 smaller fishing vessels. Panama, Namibia and Costa Rica have all also committed publicly to publish their own VMS data on the Global Fishing Watch map. All of this 'free' data provides a great boon to academics and other fisheries management experts looking to make a difference.

A middle ground is better data sharing between countries (rather than making it all completely open). In 2017, eight East African coastal countries came together to share their own data between themselves to help tackle illegal fishing. Non-profit organization FISH-i Africa, funded by the Pew Charitable Trusts, supported the nations to share vessel data in real time. By also offering access to satellite tracking expertise, the assembled task force enabled the authorities to identify and act together against large-scale IUUF. An early example of success featured progress in a $1.6 million fraud case that could only possibly have been solved by the pooling of information, but the potential is clearly much broader.

Conclusion

Just as the world relies more on fish than it ever has before, it is also awash with ocean data in unprecedented levels. The convergence of the smallsat revolution providing remarkable new quantities of Earth observation data with the continued rise of computer processing power and new analytics capabilities all lead to an entirely new arena of fisheries management, especially as it applies to countering illegal, unreported and unregulated fishing.

The continued progress of open data will hopefully put even more wind into the sails of related initiatives, as coastguards and fisheries ministries within governments come to the realization that their data can be far more powerful when made freely available, so that it can be analysed by many more experts, and combined where necessary using expertise and processing power not always readily available in the public sector.

Thankfully, powerful case studies of positive usage of open data are emerging in several domains, encouraged by NGOs and aid agencies promoting the benefits of open data, and many academics and start-ups are enthusiastically embracing this information flow to illustrate the value that can be derived. With the World Bank and the United Nations employing highly experienced data scientists to run their own open data initiatives, and organizations such as Orange and the Vodafone Foundation experimenting with the safe sharing of their proprietary data, the signs are encouraging for many more positive examples in the coming years.

Such case studies can provide comfort to the owners of closed information that the open data movement is indeed worthy of attention, and not an overhyped flash-in-the-pan, as many Big Data initiatives can be regarded. Innovative organizations such as OceanMind, Global Fishing Watch and FISH-i Africa are also making this journey easier for governments as they incorporate as non-profit technology companies. By making a clear statement in this way that their core focus is on social and environmental benefits rather than commercial gain, this may give added comfort to governments and related agencies that their data will be put to the uses it intends.

Aligned to other encouraging initiatives that utilize the best aspects of autonomy, AI and even the intensity of light to support sustainability, the ability to foster new partnerships around data and to extract actionable intelligence from it looks set to draw the net much tighter around those who seek to plunder the seas and oceans for their own illicit gain.

References

Fao.org (2018) Is the planet approaching 'peak fish'? Not so fast, study says. www.fao.org/news/story/en/item/1144274/icode/ (archived at https://perma.cc/P3GP-3CLU)

Stafford, R (2016) How overfishing and shark-finning could increase the pace of climate change. https://theconversation.com/how-overfishing-and-shark-finning-could-increase-the-pace-of-climate-change-67664 (archived at https://perma.cc/PVK6-2BM3)

11

Subsea monitoring

Shining a light on the mysteries of the deep

Conducted in a mix of Spanish and English, the atmosphere of the meeting in the Buenos Aires headquarters of the Armada Argentine in August 2018 was understandably subdued, but was also professional and purposeful. After some six hours of detailed discussion, the commander of the submarine flotilla, and leader of the session, was asked what the Plan B would be if the precious target on the seabed wasn't found in the recommended search area. 'We will ask you to search the same area again,' he said simply, with quiet conviction.

This was the first operational planning meeting between representatives of seabed intelligence company Ocean Infinity and the full Armada Argentina team responsible for analysing the evidence and setting the context for the search for the ARA *San Juan*, the submarine that was tragically lost at sea – with a crew of 44 – in November 2017. ARA *San Juan* was the second of two TR-1700 class submarines, a bespoke design built by Thyssen Nordseewerke, Germany, specifically for Argentina.

Once she was lost at sea, the immediate initial, sadly unsuccessful, search and rescue operation was followed by additional multi-national searches and surveys in the following weeks and months. All of this was unfortunately to no avail, so specilaist US seabed-intelligence company Ocean Infinity was eventually contracted for a dedicated, comprehensive subsea survey using leading edge marine autonomous technology.

NLA International Director Nick Lambert was privileged to be one of the project's two directors, alongside and supporting the vastly experienced Andy Sherrell, whose understanding and experience of all manner of autonomous and robotic marine systems in challenging environments is comprehensive. As is critical in such scenarios, he also has a remarkable ability to absorb vast quantities of data, which is becoming ever more important as such data flows are continually increasing in size and complexity.

If a tangible, physical sign of the Argentine Navy's commitment to the project were needed, it was there in the room. This was clearly no futile tick-box exercise. Some 40 experienced naval experts from the Argentine submarine flotilla, with skills and knowledge encompassing maritime operations to meteorology and hydrography were assembled, each bringing their own intelligence and detailed understanding of an issue that had occupied their daily working lives for the best part a year. These assembled experts were quietly and expertly marshalled by two-star officer David Burden, like Nick a rear admiral and currently the Director General of Naval Materiel within the Argentine Navy. His measured professionalism and engineering background underpinned a mastery of his brief, essential if the team were to successfully steer a course through a tricky morass of navigational evidence, a myriad of theories and, inevitably, massive external speculation about an event of great political significance, understandable emotion and no little global interest. His empathy for the plight and needs of the *San Juan* families was palpable and impressive.

Between them, the meeting hosts briefed the assembled experts on the details of the troubling case. They firmly believed that the *San Juan* was somewhere within a 12 by 18 mile area, an assertion that was ultimately proven to be correct. This assessment was based on a range of factors, including the submarine's last known GPS position; its planned and authorized navigational track, heading and speed (known as the SUBNOTE in submariner parlance); comparisons with Iridium satellite communications positions (the satellite providing primary communications for the boat); a water column current in that area (northerly at two knots); and the effect of the appalling prevailing weather on the submarine's operational manoeuvring options when surfaced, at periscope depth or deeper. Alongside that hard and finite data, the project team also carefully considered submarine standard operating procedures and human factors to develop as comprehensive as possible situational awareness of the prevailing operating parameters. ARA *San Juan*'s crew were experienced submariners who had been on patrol for several weeks; they knew the performance of their boat and they had worked as a team. The captain was known as a highly professional operator, certainly not given to erratic decisions or to wander off a prescribed course.

The Argentine Navy team were open and collaborative; they provided all the available information and ably fielded as many follow-up questions and requests as were put their way. Departing Buenos Aires, the team's hard work really began, systematically assimilating and understanding the navigational evidence, establishing likely operational scenarios, drawing on the advice of numerous international experts and planning the 60 days of detailed survey activity that constituted the project. Numerous other agencies also lent support, including the UK's Defence Attaché to Buenos Aires, the Royal Navy's Submarine Parachute Group, the US Navy, and experts at the Comprehensive Test Ban Treaty Organization and the UK National Data Centre.

Advanced AUVs

Having attended so many conferences and trade shows over many years, the authors can attest that the term 'state-of-the-art' can be applied a little too liberally in today's tech-hungry world. However, Ocean Infinity's flagship multi-purpose offshore vessel *Seabed Constructor* and her fleet of autonomous underwater vehicles (AUVs) without doubt more than warrant the epithet.

Understanding the bathymetry of the seabed and the depth of the water column is a blend of science and art. Measuring the depth of water using lead and line dates back to ancient civilization and even today – despite the best efforts of the kind of satellite-derived bathymetry providers featured in Chapter 8 – many, many of the maritime charts used by mariners rely on lead line surveys of considerable age. The stark fact that only 10–15 per cent of our ocean floor has been mapped using modern sonar technologies simply goes to underline just how difficult it is to operate in the often turbulent deep water environments of the world's seas and oceans, a task now facing the ARA *San Juan* search and recovery team.

Recent developments to monitor the seabed have progressed through single beam echosounders, multi-beam echosounders and towed side scan sonars complemented by Remotely-Operated Vehicles (ROVs) and Autonomous Underwater Vehicles (AUVs). The *Seabed Constructor* is the current advanced logical conclusion of that development – a vessel specifically designed to support the deployment of a fleet of AUVs (a total of five were used in the search for ARA *San Juan*) which, working in tandem, can traverse and collect data 900 line km a day. This rate of coverage would have simply been unthinkable even only a decade ago.

The ability of the latest AUVs to reach the lower depths of the ocean is only useful if they can deploy highly capable sensors. Thankfully, the industry has also focused on ruggedizing and miniaturizing sensor suites specifically for this purpose. This means that the hard-working Ocean Infinity AUVs are equipped with a seriously impressive payload: EM2040 multi-beam echosounders; Edgetech sidescan sonars and sub-bottom profilers; high definition colour cameras; conductivity, temperature and depth sensors; and self-compensating magnetometers to accurately record the magnetic field of the various seabed features. As even the most hardened tech evangelist may have struggled to predict the ferocious speed at which the smartphone has incorporated such a broad set of sensors as it has (from music player to video camera to accelerometer), the latest AUVs are indeed similar small miracles of engineering and data-capture power.

Establishing the search rhythm

The ARA *San Juan* subsea search started on 6 September 2018 and followed a regular rhythm. Starting in areas of high confidence before expanding incrementally, the *Seabed Constructor* methodically proceeded to the designated area of interest, deployed the AUVs to do their work in the deep seabed canyons and rivers that border the Argentine continental shelf, recovered them and moved on to the next area of interest. Collected data would be analysed on board and shared with a variety of experts in the USA via satellite communications for comparison and corroboration. Occasional port calls for re-supply and crew change enabled operational pauses and a chance to reflect on progress. Running alongside and supporting this operational rhythm, the multidisciplinary team constantly questioned the analysis and logic for the planned search areas, poring over the unfolding shape of the seabed, identifying geological features, discarded fishing nets and long lost ship wrecks, constantly challenging assumptions, interpreting and reinterpreting, and corroborating vast quantities of data new and old.

Despite being driven by new technology, the involvement of a diverse range of additional expertise was crucial to the operation in such a complex and challenging seaspace. The broader team comprised world-leading experts in the intricacies of position, navigation and timing; satellite communications; subsurface acoustic anomalies; submarine and surface ship operations; together with engineers and naval architects. Although scattered around several countries, they instinctively worked together, reviewing the evidence and comparing the possible *San Juan* scenarios with other lost submarines, including the implosions of the USS *Thresher* and the USS *Scorpion*, which were sadly lost at sea in 1963 and 1968 respectively. The sense of purpose, unstinting support, collaboration and teamwork – invariably provided in a voluntary capacity – were inspiring and humbling. While it is easy to believe those who say we live in a fractured world, such pessimists would have taken pause had they witnessed this project.

This spirit of collaboration was, of course, extended to the understandably distraught but resolute families of the ARA *San Juan*'s lost crew. Oliver Plunkett, the visionary CEO of Ocean Infinity, took the time to personally brief the families before the operation commenced and they were dutifully kept abreast of significant developments, as a matter of course. Family members were also invited to join the crew of the *Seabed Constructor*, and daily updates were disseminated further by a willing, energetic and supportive set of citizen journalists via their own social media channels, who in turn received daily encouragement from their enthralled and supportive online followers. While balanced reporting was very much the order of the day throughout, nobody was ever in any doubt about how much this project meant to so many.

Continual challenge

The continual challenge of the data and the operation's logic took Ocean Infinity's AUVs to some interesting places as the extremities of the available evidence were explored, factoring in the likely performance of the submarine and the constraints on her operations. Arguably, the key piece of evidence was the anomalous acoustic event detected by the Comprehensive Nuclear Test Ban Treaty Organization (CTBTO). This international agency operates a global network of acoustic sounders, monitoring the oceans for underwater seismic events such as nuclear explosions; it can also (and, in this case, did) pick up other significant natural and manmade acoustic events, such as an imploding submarine. At 13:51 GMT on 15 November 2017, CTBTO Hydroacoustic stations HA10 (Ascension Island) and HA04 (Crozet) had detected a signal from an underwater impulsive event that occurred in the vicinity of the last known position of the *San Juan*. The data was of course made available to the Argentinian Authorities.

Initially the cross-referenced CTBTO acoustic bearings and the estimated depth shaped the high probability search area to the east of the continental shelf but with an extensive east–west positional ellipse. While this ellipse compared well with the estimated track of the submarine, her last known GPS positions and the Iridium footprints, the original Argentine, US and UK search planners needed to further refine the east–west component. Their imaginative solution was to initiate a controlled depth charge explosion in a known location for detection by the CTBTO network. This simple exercise enabled extensive refinement of many complex calculations, reducing the ellipse error from around 180 miles to a confidence-boosting 20 miles or so.

Ultimately, the CTBTO data proved to be extraordinarily accurate. The wreckage of the stricken *San Juan* was finally located at around 11pm local time on 16 November 2018, one year and one day after her loss. Over the 60-day period of the intensive search, Ocean Infinity had deployed five-side scan sonar equipped Hugin AUVs on 150 missions. They completed a total of 5,000 survey hours, covering 30,000-line kilometres over an area of 21,000 square kilometres – again, an astonishing return in the time. The wreckage was finally located on a small ridge amongst geological features in a ravine in 920 metres of water, approximately 600 kilometres east of the Patagonian city of Comodoro Rivadavia. Her final resting place was within 11 miles of the positions estimated by the Argentine, US and UK experts who oversaw the original search in November and December 2017.

Thus, right at the end of the contracted 60 day survey and as Seabed Constructor was about to depart for South Africa and future tasking, it was time to deploy the ROVs to confirm the AUV data with photographs of the debris field and wreckage. The team's knowledge of the *Scorpion* and *Thresher* wrecks suggested that the most identifiable characteristics of *San Juan* would be her propeller and stern section, the conning tower (a raised platform sometimes called the fin) and her distinctively broad, flat bow. A few hours before the entire operation was scheduled to finish, the startlingly high-quality images came back. There could be no mistaking the propeller and shaft, nor the distinctive shape of the conning tower and the already weed-encrusted bow section with torpedo tubes. The ARA *San Juan* had been found, providing some comfort to the families, friends and colleagues of the fallen.

Needles in a watery haystack

While cases of the severity of the ARA *San Juan* are thankfully rare, the seafloor is littered with sunken ships. We covered hydrography and bathymetry in some detail in Chapter 8, but aside from general understanding of the shape and movements of the seabed, many governments, NGOs and companies have a strong interest, and quite often an urgent need, to find specific items in the vast expanse of the oceans' murky depths, some of which have evaded detection for considerable passages of time.

There were 94 total ship losses in the commercial shipping fleet in 2017, the lowest for 10 years according to data from insurer Allianz. This is quite a small number when one considers that the commercial fleet contains about 40,000 vessels, and that they often operate in difficult weather conditions in congested seaspace. Apart from the risk of sinking or crashing, though – incidences of which are fortunately very rare – expensive and classified items of equipment can fall off vessels at sea all the time, either by accident or through deliberate jettison to achieve a short-term, perhaps emergency, aim. Ships can lose equipment and munitions in heavy weather, while replenishing at sea, when jettisoning topweight after an incident involving stability, to move them out of the way of fire and other hazards, during firing trials of existing or more importantly experimental/developmental armament, and in random accidents such as collision and grounding.

When ships sink there is often enough debris, and AIS, communications or satellite information to provide fast clues to location. That said, ships on clandestine operations may have disabled automatic ship–shore position reporting and satellite communications location data, meaning that identifying the location is more problematic. Submarines lose equipment mainly through the release of weapons which then fail to run or fly, and sink into the sea – this might be because the flight or guidance system failed, or the self-destruct system failed, or, in the case or torpedoes, the homing and 'bottom capture' sequence did not work, leaving a live or exercise weapon on the seabed.

As the unfortunate ARA *San Juan* case brings home in stark relief, submarines have significantly elevated levels of risk. Forty-one nations have currently acquired submarine capability, and submarines have (very occasionally) lost entire towed array systems, propellers and rudders. They can collide on the surface and underwater and may lose valuable/classified equipment in those incidents. When submarines or warships sink, their entire equipment is available, theoretically, to a third party salvor and their agents, so governments are always keen to retrieve them, especially when lost assets include confidential or classified information or high-powered military technical equipment that could be copied by unfriendly forces.

All eyes on flight MH370

One of the major global news stories of recent years that relates to the ocean revolved (and sadly still revolves) around the unexplained loss of flight MH370. The disappearance of this Malaysia Airlines flight, en route from Kuala Lumpur to Beijing on 8 March 2014, led to the largest and most expensive search in aviation history. Despite vast effort, notably in the hostile South Indian Ocean, nothing was found until July 2015, when a battered aircraft wing part washed up on Reunion Island. French officials soon confirmed that the debris was indeed from MH370. The main body has yet to be found at the time of writing, though on the fifth anniversary of the crash Malaysian officials suggested that they were considering re-opening endeavours. The MH370 case brought unprecedented attention to the challenges of locating aircraft that have crashed over water, even using subsea monitoring devices.

Thankfully, major civilian airline crashes over water are again rare (though at least 15 flight data recorders lost at sea have never been located or recovered) and becoming more so. That is because, first and foremost, commercial aviation is a very safe business. Research from Boeing shows a consistently low and decreasing accident rate. In the last 14 years there have only been 12 commercial airline crashes into water. In 2017, in fact, there were no civil airline crashes, on land or at sea.

When crashes do happen, the vast majority of them occur over land or very close to the shore during the take-off or landing phase of a flight. Nearly half of all worldwide commercial jet airplane accidents during the 10-year period from 2004 to 2013 occurred during the final approach or landing phases of flight, according to statistics published in September 2018 by Boeing Commercial Airplanes. Of the 72 fatal accidents recorded during the period, 22 per cent (16 accidents) occurred during the final approach phase and 25 per cent (18) during landing. Deep water crashes are, by contrast, a relative rarity.

So, while subsea monitoring specialists with state-of-the art technology will remain vigilant in relation to aircraft dropping into the sea, they should thankfully not be kept overly busy with those kinds of missions. There is plenty of other seafloor material to keep them occupied, though, some of which has been undiscovered for some considerable time.

Looking further back, many centuries of ocean travel have left various parts of the seafloor littered with wrecks. It is known, for example, that around 2,000 ships were lost in the Gulf of Mexico between 1625 and 1951. Up to 1,300 undiscovered wrecks are thought to lie off the western coast of Australia. Further clusters align to ancient/historic trade routes or known major battle sites, for example Aboukir Bay, Goodwin Sands and the Dalmatian Coast, which was one of the world's most intensive shipping routes in the second and first century BC.

One of the main legacies of the devastation of the Second World War is the vast number of sunken ships that lie on the ocean floor. As many of them were warships, they also contain a great deal of unexploded ordnance, so in many ways can be viewed as significant accidents waiting to happen. The greatest density of these potentially polluting wrecks is to be found in the south Asian Pacific where 2,737 sunken ships and submarines rest forlornly on the seabed. Second on the list is the north-west Atlantic with 1,393 sunken vessels, followed by the north-west Pacific with 1,152 and the north-east Atlantic with 786. All these areas, unsurprisingly, were key battlegrounds.

While such wrecks obviously hold great interest for archaeologists and war historians, they also present a more pressing need, and one that will grow more urgent over time. Many of these vessels sank with their hulls relatively intact, which means that they still contain a huge amount of heavy fuel.

While they will undergo natural degradation, another man-made issue related to the ongoing integrity of such wrecks came to light recently. In March 2019, new academic research findings revealed that the 4.9 million barrels of oil that were spilled into the Gulf of Mexico during the 2010 Deepwater Horizon catastrophe were having an alarming effect on wrecks in the area. The report, published in the journal *Frontiers in Marine Science*, explained how the extra corrosive elements introduced by the oil were causing Nazi submarines to disintegrate at a far faster rate than had been observed previously (Mugge et al, 2019). The oil created a more supportive environment for certain types of bacteria, which stick to ships' hulls and at the same time unfortunately release a corrosive biological byproduct that has the potential to degrade metal. The hull of one wreck, the German submarine *U-166*, was sunk by a depth charge in 1942 and identified by acoustic imagery provided by an AUV in 2001, some 45 miles southeast of the mouth of the Mississippi River, at a depth of almost 5,000 feet.

There are estimated to be some 8,500 potentially polluting wrecks in total, 6,735 of which are from the Second World War period or before. Four areas of interest are worthy of closer attention. Starting in the Pacific region, 3,100 wrecks have been located, predominantly US or Japanese in origin. The majority were conscripted merchant ships from Japan serving as cargo vessels. More worryingly, approximately 10 per cent of all these lost wrecks were oil tankers. As in other areas, the marine environment has slowly corroded the steel of their hulls over the past seven decades, compromising their interior compartments and their ordnance. In some known cases, less than 40 per cent of the original material remains, so management decisions and subsequent interventions are urgently required to avoid foreseeable environmental disasters.

Raising the profile of dangerous wrecks

Small Island Developing States (SIDS) continue to endure the greatest impact from Second World War wrecks. SIDS, of course, potentially have the most to lose (as a larger proportion of their economies rely on the sea), but often the least amount of money to spend to support the location of potentially polluting ships on the seabed or to undertake salvage of known troublesome wreck sites. Currently, there are two large development and conservation initiatives at work in the Pacific region attempting to address this problem and keep the issue of potentially polluting Second World War wrecks in the public eye so the need for urgent remedial actions cannot so easily be overlooked.

Both the Coral Triangle Initiative (CTI) and the Secretariat of the Pacific Regional Environment Programme (SPREP) have undertaken detailed analysis of wrecks in the western Pacific. SPREP, comprising 21 nations (including the United States and France), is the largest regional organization focused on environmental issues. A total of 656 sunk ships have been identified in its waters. The CTI, formed of the Philippines, Malaysia, Timor-Leste, Papua New Guinea and the Solomon Islands, has identified 997 sunken ships in its area of interest. The waters of the Philippines provide the most populous wreck-hunting ground, followed by Indonesia. Of particular note in this area is the prevalence of disturbance events; tsunamis, earthquakes and volcanoes are all common in the tectonic 'Ring of Fire' region. Each such event has the potential to disturb wrecks, hastening their collapse, so the serious effects even of incidents in the past few months may be happening right now but not yet detected.

Moving across to the North Atlantic, where the US Coast Guard is responsible for local area contingency plans in case of a spill, the National Oceanic and Atmospheric Administration (NOAA) has led the way in this area. It gathers data on US wrecks and has undertaken sophisticated modelling to arrive at a published ranking list of potential oil spill incidents emanating from the wrecks in its waters. NOAA has 87 ships on its watch list, and not all exact locations are known. The list – accompanied by detailed analysis of each wreck site – includes vessels known to have been lost but whose location has not yet been identified; wrecks whose locations are known but are not considered to present a significant threat; those that may cause harm and require monitoring; and those that are known to have suffered severe structural damage and are therefore considered to be a high priority for intervention, such as the *Esso Gettysburg*.

Canada is adopting a similar line to its North American neighbour. In October 2017, to mark the one-year anniversary of the country's $1.5-billion Oceans Protection Plan, the Government of Canada announced the introduction of the Wrecked, Abandoned or Hazardous Vessels Act (Bill C-64) in Parliament. The National Strategy to Address Abandoned and Wrecked Vessels seeks to prevent the occurrence of new problem vessels and make progress in cleaning up existing problem vessels. Specifically, the activity aims to 'empower the Government of Canada to take proactive action on hazardous vessels before they become more costly to Canadians'.

In May 2017, Transport Canada also launched a five-year, $6.85 million Abandoned Boats Program, which provides funding support for the removal and disposal of hazardous small boats. To further support its drive to clean up the ocean floor of dangerous wrecks, in January 2019 the Canadian Government awarded a contract to develop a risk assessment methodology related to hundreds of vessels of concern (abandoned, wrecked or dilapidated vessels) in Canadian waters or on Crown land. The contract was valued at over five hundred thousand dollars and was awarded to UK company London Offshore Consultants. In time, this will lead to a list like NOAA's.

The task will then, of course, relate to what to do with such information. How can remedial action be taken in a timely manner, and not left until situations become critical? What can technology do to help meet that need?

New possibilities

The fact that it took so long to find the ARA *San Juan*, that many re-surveys were required and that she was found so close to the estimated position demonstrates that, notwithstanding the most sophisticated subsea technologies, our seas and oceans are remote, challenging and difficult to survey. A reflection of the critical factors that ensured mission success underlines the ongoing need for man and machine to work together to obtain best value. The operational quality of the ship and the AUVs deployed are almost a given but still worth repeating to underline their importance. The HUGIN AUVs were able to operate between depths of 5 metres right down to 6,000 metres. As they are autonomous, they don't have to be tethered to the mother ship and so can use their powerful suite of high definition sensors to very quickly provide large and wide coverage of the survey area, acquiring a huge amount of high-quality data incredibly quickly. The amount and quality of evidence provided by the Armada Argentine enabled coherent, detailed planning and high confidence and the continued flexibility of the team (in its broadest sense) enabled the operation to adapt to an ever-evolving situation.

As well as providing comfort to the families of the lost submariners, the successful mission locating the ARA *San Juan* also showcased the new levels of subsea monitoring capability made possible by technological advances. This is encouraging because there is much yet to find on the mass expanses of the ocean floor.

As we have seen, the seafloor is littered with masses of debris – some centuries old, some added more recently. Overlooking the plastic waste that adds new layers to the seabed in various parts of the world (described in greater detail in Chapter 9), the ocean has collected many more wrecked vessels before the ARA *San Juan* fell to the bottom and will continue to do so in the years ahead. So further technological progress will be required to help meet the growing need.

The 'system of systems'

As part of the authors' ongoing market research into the potential of unmanned and autonomous systems, we hear again and again of the desire for surface and subsea vessels to work together in a 'system of systems' that delivers enhanced capability.

In the first instance, this can relate to navigation and the ability to deliver a safe and secure relay of expensive subsea data. More than anything, subsea surveyors utilizing remotely-operated underwater vehicles (ROVs) and autonomous underwater vehicles (AUVs) are terrified of losing either the data they have collected, or the devices themselves. Many are therefore interested in unmanned surface vehicles (USVs) as navigation aids for AUVs. As evidenced in the ARA *San Juan* mission, some AUVs are very long range and can go up to 6,000 metres deep. Operators can put the AUV in the water, but many have told us that they would like more assurance as to where it is at any point in time. Deployment of a linked unmanned surface vessel would enable 'over the horizon' navigation – to follow the AUV along and let operators know exactly where it is.

One of the biggest challenges of subsea monitoring relates to the collection of data subsurface and then getting it back to shore safely. If a USV can provide real-time, satellite-enabled communications, operators can reduce the potential risk of losing their AUV and all the data it has collected. Surface vessels would then act as a communications platform, a data link to underwater vessels undertaking survey operations, to get the data back to shore – akin to an autonomous 'mother ship'. Transmitting data in this way as operations are under way, rather than recovering the vessel and then looking at all the data, also reduces the potential risk of no useful data being collected at all, for example if the data-recording sensors and instruments weren't set up properly and weren't recording accurately or at all.

One of the sometimes-overlooked benefits of unmanned and autonomous vessels is just how easy they are to launch. One subsea survey manager who spoke to us recently commented that with manned surveys they must be heavily focused on the logistics rather than the data itself. Eighty per cent of their effort, he said, focuses on putting the boat in the water, which necessarily takes valuable resource away from what should really be the focus – the analysis of the expensively collected data.

There is a clear market pull, then, for an easily launched autonomous system of systems to collect data in the deep sea and get it to a companion device on the surface which then, via satellite, relays it to shore in near real time. Ease of deployment and recovery is key – to be light touch on the operations side, to allow all the focus to be on the analytics.

Safeguarding the world's data

Another aspect of the seafloor that requires close, continual monitoring, and often swift intervention, relates to the extensive network of cables that criss-cross between continents, carrying the world's power, data and communications. There are over 1.2 million kilometres of submarine cables in service globally, and a staggering 97 per cent of all intercontinental data is carried via such underwater infrastructure.

The first transatlantic cable was completed way back in 1858, when a congratulatory message from Britain's Queen Victoria conveyed to US President James Buchanan opened up a new era of communication between Britain and the United States via the Atlantic Telegraph. Queen Victoria's message took a total of 16 hours to transmit over the seven copper wires, but this was a marked improvement on the approximately 12 days it would previously have taken to deliver the royal message across the Atlantic, initially by steam-powered ship and subsequently utilizing the well-established land telegraph.

President Buchanan's response was suitably grand in recognition of this landmark, writing as he did: 'May the Atlantic Telegraph, under the blessing of Heaven, prove to be a bond of perpetual peace and friendship between the kindred nations, and an instrument designed by Divine Providence to diffuse religion, civilization, liberty and law throughout the world.'

The news that such an innovation had the potential to revolutionize communication over the long stretches of water between Europe and the United States was received with genuinely wild excitement and enthusiasm on both sides of the Atlantic. New York City, for example, celebrated the occasion with everything from a 100-gun salute to a parade and torchlight procession. So the razzmatazz that surrounds new tech launches may have a longer heritage than has previously been realized.

However, soon after its auspicious start, the much-lauded transatlantic cable was beset with technical difficulties. It finally failed completely the month after its original launch when, it was reported, a technician overloaded the system when attempting to achieve a greater speed of transmission. Despite the promise shown that the innovative technology *could* work, investor confidence waned, meaning that it would be another eight years before the project was resurrected in a more stable form. So, again, maybe some of the 'valley of death' challenges facing 21st century technology innovators is not so new either.

Since those early days two centuries ago, there has been a steadily increasing desire for greater speed of data delivery across the ocean floor. Coming back to the current day, Google announced in early 2019 that its planned Dunant cable (named after Henry Dunant, the founder of the Red Cross) will transfer data through 12 pairs of fibre-optic strands across 3,977 miles of seabed from the United States to France at a rate of 250 terabits of data per second. By rough calculation this would mean that, when operating at full capacity, the Dunant transatlantic cable will be able to transfer approximately 9.9 million ultra-high-definition movies *a second*. Some progress, then, from the handful of words originally transmitted by Queen Victoria.

The desire for such speed is understandably also matched with the equally pressing need to ensure that the integrity and continual operation of these increasingly essential subsea cables so that the systems, services and often entire businesses that have rapidly come to rely on them are not compromised. While modern cabling is robust, no system is fool-proof, and subsea cables do fail; on average, there are over a hundred subsea cable breaks each year. Approximately 70 per cent of all of these cable faults are caused by fishing and anchoring activities (Southeast Asia has comparatively high incidences of cable failure due to the high intensity of fishing activity in its waters) and around 12 per cent are caused by natural hazards, for example due to earthquakes or significant damaging abrasion caused by strong currents.

The fragility of the connected world

Cables breaking can have extremely damaging consequences, as evidenced by the unfortunate case in January 2019 when the 827-kilometre subsea cable that connects the Kingdom of Tonga to Fiji failed completely. In operation since August 2013, this subsea cable system was jointly funded by the Asian Development Bank and the World Bank, underlining the growing importance of such endeavours to island communities.

When the cable failed on a Sunday evening, Tonga lost all mobile phone and internet connection to the outside world, with widespread consequences for both the business and personal sectors in the Polynesian archipelago. Not surprisingly, the government worked frantically but with purpose to find a solution. In the first instance they worked with satellite providers to increase connectivity while the break in the cable itself could be located and repaired. With reduced service, however, and looking to protect the availability and speed of other internet-powered services, the Government of Tonga explored the possibility of blocking access to social media websites, which they revealed accounts for approximately 80 per cent of the country's international traffic. As they grappled with the full extent of the implications of a life without internet and cell phone connectivity, Paula Piveni Piukala, Director of Tonga Cable, said that the unexpected subsea cable failure was 'a wakeup call for small countries like us in the Pacific'.

Again, as in so many Blue Economy domains, it may be a case of robots to the rescue. Once the location of a cable fault has been identified (itself a not inconsiderable task), the responsible company can send out a bespoke cable ship that will carry on board a few miles of fresh fibre-optic lines to make the repairs. How that is achieved depends on how deep the offending cable is. If the faulty part of the cable is less than 6,500 feet down, the crew will utilize a submersible, tank-like robot that can move around on the seafloor. Such robots can be guided towards the identified area of the problem by sending signals to the cable. When the robot reaches the right place, it will clasp the cable and then cut out the section that needs to be replaced. The loose ends are then pulled back up to the ship.

In depths greater than 6,500 feet, very high water pressure makes the operation of such robots impossible. In these situations, technicians aboard the cable ship instead use a hook on a (very) long wire to pull up the cable from the seafloor. Much like the robots, this device uses a mechanical cutting and gripping device that can split the offending cable section on both sides of the break and then drag the loose ends to the surface where it is repaired at sea by splicing the fibres and attaching the new section with specialized adhesives.

The world's insatiable hunger for data means that the submarine cable systems market is predicted to grow rapidly in the coming years, reaching a headline value of £15.9 billion by 2023, meaning that it will grow at a compound annual growth rate of 12.25 per cent from 2018 to 2023 (Tan, 2018).

While the smartphone revolution has in many cross-cutting ways kick-started the exponential growth of data services, it is the continued rise of sectors such as the Internet of Things that is really supercharging them. Machine-to-machine traffic has multiplied fivefold in the past five years, to a point where it now accounts for approximately 80 per cent of all subsea traffic flowing through the data cables across the Atlantic.

The largest change in this market in recent years is the entry of major social media and cloud computing companies. Whereas before they would rely on cable services provided by others, the highly ambitious growth plans of cash-rich corporate behemoths like Google, Facebook, Microsoft and Amazon have drawn them into serious infrastructure development, becoming major investors in new subsea capacity. The Dunant cable is just one of many new lines in development.

Since 2010, Google alone has invested in 14 new subsea cable systems and both Facebook and Google are investors in the planned Pacific Light Cable Network that will link a landing point in California to several in Southeast Asia (the Philippines, Hong Kong, Taiwan). By taking responsibility for this subsea infrastructure, the big players take control in greater measure of the integrity of data flows into their giant data centre campuses, on which their entire business operations rely. In fact, Google has commented publicly that directly connecting its data centres without having to consider the needs of other cable consortium partners (which is the traditional model of funding such infrastructure) is one of the main drivers of their investment in this area. Put simply, they can lay their cables so that they service their shore-based data centres as efficiently as possible and take sole ownership of all of the bandwidth within them for the 15–25 years that such cables usually last.

Getting power back to shore

Another growing market area for subsea cables is the offshore renewable energy market. As new offshore wind, wave and tidal systems continue to proliferate (all of which are covered in full in Chapter 4) more subsea cables are required to transfer their naturally produced energy back to shore and into the grid.

A report launched in January 2017 underlined how important subsea cable transfer is to the overall operational effectiveness of the offshore renewables market, as German insurance company Genillard and Co reported that, at the time, some 4,600 subsea cables had been laid to service offshore wind farms. Within this sector, they reported that, on average, 10 cable failures were reported a year, with an average cable downtime of 100 days. The eye-catching headline figure, however, was that just over £350 million had been incurred in claims related to subsea cable losses in the offshore renewables market in the previous seven years.

Attempting to look deeper into and more proactively at the problem, in September 2018 the UK's Offshore Renewables Catapult (a technology innovation and research centre) produced a report that looked at the ability to predict the failure of subsea cables connecting offshore renewables infrastructure. Acknowledging that the problem was especially prevalent in the offshore renewables industry due to the increasingly dynamic nature of the waters as, for example, floating wind installations are sited further and further out to sea, the paper outlined the early stages of a research project that will attempt in the end to move towards predictive modelling of floating wind subsea cable breaks.

As with all data-fuelled predictive modelling projects, it will start in the first instance by focusing on descriptive analytics, in this case studying how dynamic subsea cables servicing offshore floating wind farms interact with their environment. This includes waves and currents, as you would expect, but the additional complication of the movements of the floating platform will not yet have been studied in as fine detail in a still-emerging market.

Within the broader subsea cables market, others are also doubling down to understand better cable failure rates and push towards greater stability with proactive measures. In January 2019, Australian telecommunications company Telstra – which runs cables across the difficult Asia Pacific region – proudly announced that it had overseen a 30 per cent reduction in service impacts on its subsea cables network. Telstra's fibre-optic subsea cable network runs for more than 400,000 kilometres under the ocean floor. Laid end to end, it would circle the world nearly 10 times, so a 30 per cent reduction in incidents across such infrastructure is a considerable gain.

They achieved the outcome of fewer cables being cut and damaged through three strands of activity. First, they utilized the power of a ship's automatic identification system, which transmits a vessel's position, speed and direction. As one of the major threats to subsea cables is commercial vessels damaging cables by dropping heavy anchors on them, the Telstra team would monitor the movement of ships in the company's cabled areas, and proactively contact the vessel if they came uncomfortably close, so that they would not stop and drop anchor. They reported that by deploying this system alone – using already available technology – they had reduced the number of incidents caused by commercial vessels by approximately 20 per cent. They did note, however, that in dangerous sea states it is not always possible for ships to delay dropping anchor, as they may require additional stability in heavy weather.

Second, to help to counter the significant threat from bottom-trawling fishermen, Telstra simply led community engagement exercises with groups that ranged from fishing unions and collectives to local fishing villages. In several informal sessions, the Telstra team shared the locations of their cables on charts and encouraged contact if the fishermen were ever to snag a cable accidentally.

Third, Telstra invested in better use of available data to help with planning the routes of cables. As we have seen repeatedly throughout this book, more and more data sources about the ocean are being collected all the time, all of which provide greater sources of accuracy for planning. Companies such as Telstra can now use multiple satellite and bathymetric data sources to enable them to plan cable routes that will help them to avoid congested shipping routes and heavy areas of fishing activity, which provide the greatest risk to their precious seabed infrastructure.

Conclusion

Many of the chapters within this book look to the skies to highlight ground-breaking satellite-enabled technologies that are showing potential to achieve real value within several Blue Economy sectors. However, various industry moves point to ever more activity than before taking place on the seafloor itself, so we must continue to look down as well as up for innovative approaches.

Many shipwrecks that have been resting quietly on the seafloor for several decades are reaching the point where their potentially damaging cargo can no longer be ignored. It is clear that Second World War wrecks on the seafloor present a significant risk to the environmental safety of the marine and coastal environments of many parts of the world, with the potential to significantly impact nations both rich and poor. Many of these wreck sites can only be viewed as major accidents waiting to happen, so further investigation and salvage operations are becoming increasingly essential. NOAA's sophisticated modelling of the risk factors of such sites is to be welcomed and will help to align available resource to the most pressing areas of need in the next few years.

As seaborne traffic increases further across several sectors, many more precious items may fall to the seabed that need to be retrieved urgently. Such cases will range from retrieving commercial flight data recorders that need to be accessed to understand why disasters have occurred over water to highly sophisticated military equipment that governments do not want to fall into the hands of others.

The world's insatiable demand for data – and the commercial desire of social media companies to service that demand in as controlled a way as possible – will drive the installation of many more high bandwidth subsea cables in the years to come.

All these issues create a greater demand for capacity to reach the lower depths of the oceans as quickly, efficiently and affordably as possible – to save lives and protect commercial and security interests. As the Tongan Government recently learned the hard way, it is not just avid social media users that now rely on the integrity of subsea cables; in the connected world, very few parts of civil society – from security operations to healthcare systems – can now be said to truly stand alone from the need for internet connectivity. So, again, the need to develop more sophisticated ways of tracking and correcting problems on the seafloor are becoming more urgent.

Autonomous and remotely operated vessels are yet again proving their worth in such harsh environments, and the proactive and preventative work such as that being undertaken by the UK's Offshore Renewables Catapult, to study the behaviour of dynamic subsea cables that serve floating offshore wind so as to be able to predict where such cables may fail, provides a new Blue Economy focal point for ever more powerful data analytics capabilities. Of course, with the world's largest data companies now becoming the major investors in the installation of subsea cables, it is to be expected that they, too, will bring greater analytical heft to bear in similar endeavours. This in turn can be aligned to the intelligent use of available marine and maritime data for safe and sensible cable routing in the first place, the benefits of which companies such as Telstra are beginning to realize to reduce significant incidences of outages.

Obviously, when reflecting on the tragic case of the ARA *San Juan*, swift deployment of suitably capable subsea monitoring assets could also save many lives. While in this case it was not possible to find the vessel on the seafloor in time to save the lives of the crew, the eventual locating of the stricken submarine does provide some encouragement for the potential of future operations. Many forward-thinking navies around the world are actively looking to learn from how the combination of newly available, leading edge technology and expertise, galvanized by hardened tenacity and truly global collaboration, drove the ARA *San Juan* project to eventual success. Such open-mindedness bodes well for the future of search, rescue and recovery operations challenged by the magnitude, majesty and challenges of the world's seas and oceans.

References

Mugge, R, Brock, M, Salerno, J, Damour, M, Church, R, Lee, J and Hamdan, L (2019) Deep-sea biofilms, historic shipwreck preservation and the deepwater horizon spill. www.frontiersin.org/articles/10.3389/fmars.2019.00048/full (archived at https://perma.cc/F4EA-MPVW)

Tan, A (2018) Subsea cable trends and opportunities beyond 2018. www.telecomasia.net/content/subsea-cable-trends-and-opportunities-beyond-2018 (archived at https://perma.cc/Z7M9-8HC9)

12

Safety of life at sea

Protecting human life in the harshest of environments

As well as being a glorious haven for biodiversity, the location of billions of pounds of Blue Economy-related business and the source of daily pleasure for millions, the world's seas and oceans are undoubtedly dangerous places. The obvious and ever-present threat of drowning is exacerbated by the prevalence of essential heavy machinery, sudden and dramatic changes in weather, in some cases laxly enforced regulations and – ever more noticeable in the past few years – the increase of highly risky sea journeys by migrants seeking solace in new countries, despite the perilousness of the waters to be traversed and the gross inadequacy of the vessels in which they find themselves.

> 'There is no healing process.'

So said Florida resident Blu Stephanos, who was talking to *People* magazine in July 2018, on the third anniversary of the loss of his son. Austin Stephanos was 14 years old when he and his great friend Perry Cohen set out for a fishing expedition in the Atlantic waters off Jupiter Inlet, Palm Beach County. They never returned. Their bodies have never been found.

> 'He left on Tuesday around noon, but he never came home. I do not know how I am going to raise my two children.'

Aisha Williams was attempting to come to terms with the loss of her husband in September 2018, one of over 220 victims of Tanzania's catastrophic – in fact, worst ever – ferry disaster. Aisha had come to collect his body. Despite only having a licensed capacity of 100 passengers, eyewitnesses said that the ageing MV *Nyerere* was carrying more than 300, as well as cargo. It capsized just a few metres from the safety of the dock.

> 'It should be safety first, and money after. Some tend to have money first and safety after.'

Leonard Leblanc, President of the Fisheries Safety Association of Nova Scotia, was speaking in Canada in October 2018. A retired fisherman himself, he was addressing press as the Transportation Safety Board of Canada revealed that 2018 had been the deadliest year in over a decade for the country's commercial fishermen, with 17 perishing on board fishing vessels.

> 'When it happened, I couldn't believe it was him, that it happened to us. There's no words. No words.'

Marisa Medici was talking about her nephew, Arthur Medici. The 26-year-old was boogie boarding in Cape Cod Bay, Massachusetts when he was pulled under the water by a shark. Eventually dragged to the shore by well-wishers, Arthur was transferred to Cape Cod Hospital but soon succumbed to his injuries to become the first victim of a deadly shark attack in the area in nearly 80 years.

> 'Let's be clear: he never should have been put in this dangerous position.'

Jacqueline Smith, Maritime Coordinator for the International Transport Workers' Federation (ITF), was reflecting on the plight of Dennis Gomez Regana, a Philippine seafarer who had been killed in Southbank Quay in the Port of Dublin in November 2018 (SAFETY4SEA, 2018). Mr Regana – a crew member of the Antigua and Barbuda-registered containership MV *Francop* – had been securing a shipping container (a procedure known as lashing) when he was crushed to death by a container. Such patently dangerous work should be the exclusive preserve of dock workers familiar with the environment, not recently arrived seafarers, said the ITF.

'The victim lost his balance, slipped and fell into the sea.'

Finally, local police Chief Commander Ketut Suastika was describing the tragic events surrounding a particularly 21st century demise on the Indonesian islet of Nusa Lembongan, southeast of Bali. Speaking to *The Jakarta Post*, he described how 46-year-old Chinese tourist Liang Wanchang had lost his footing as he was attempting to take a selfie to capture the stunning ocean views seen from coastal beauty spot Devil's Tears. Widely known for the Instagram-worthy waves crashing exhilaratingly over the jutting rock formations, one wave swept the holidaymaker – and his Dutch friend Emeli Setch, who was thankfully saved – out into the unforgiving waters. Mr Wanchang was eventually declared dead on arrival at the local hospital, after the rough conditions had hampered a desperate rescue mission.

His unfortunate case is far from isolated, however, as some 259 people died while taking selfies between October 2011 and November 2017 (Bansal et al, 2018). Drowning was the top cause of selfie-related death, accounting for 27 per cent of all fatalities (transport-related deaths were second at 20 per cent).

To underline the recent acceleration of this peculiar new risk, in 2016 Mumbai police established 16 'no-selfie zones' across the Indian city's coastline following the death of a man who drowned attempting to save a girl who had fallen into the sea while taking a photo of herself. Similarly, in February 2019 the Irish Minister for Mental Health and Older People, Jim Daly, called for the introduction of 'selfie seats' in popular coastal tourist locations. His call was in response to a Trinity College student falling to his death while reportedly attempting to take a selfie atop the Cliffs of Moher in County Clare.

Such diverse incidents – and many more besides – underline the dangers that cut across all sectors of the Blue Economy, from leisure and tourism and industrial cargo shipping to transport and – as we shall see – the fishing sector and all manner of global offshore industry.

Thus, safety of life at sea is a major theme that cuts across every sector of the Blue Economy, and could reasonably have featured in every chapter of this book. While various responsible bodies globally are committed to ensuring that stringent safety regulations are in place in all marine areas – whether at work or play on the water – further innovations related to preventative, protective and emergency situations at sea are still very much in demand.

VR, wearables and robots to the rescue

As we explore in greater detail in other chapters, one very promising technological progression promoting safety of life at sea is the continued development of autonomous systems. If humans can be at risk at sea, especially within the harsher environments of the high seas, deep underwater or within the polar regions, the ideal scenario would obviously be for them not to be in these situations. This is becoming more of a realistic proposition as the benefits of autonomous sea-going vessels. As well as removing the need for human involvement in more mundane data-collection activities, this could also be extended to hazardous incidents where the full extent of the risk is not yet understood. As one port innovation manager explained to us: 'When you have a fire on a vessel, you could send a USV into the incident scene with a greater degree of safety to assess what kind of toxic gas may be present.'

Another interesting technological development within this domain actually takes place in the main on land. While any number of technologies, regulations and well thought-through policies can be deployed to enhance safety at sea, and even looking forward to a future that features more autonomous and AI-driven systems, human intervention in sea-going operations may always be necessary. Staff training, then, and more general guidance to be followed by anyone taking to water, is crucial.

The ability to deploy cutting edge digital technologies to prepare staff for sea-going activities has been progressing. In particular, the expansion of virtual reality (VR) techniques in training situations has been receiving healthy investment and interest. In January 2019, for example, Newcastle College in the North East of England announced the launch of a new VR training facility specifically targeted at the offshore wind sector. Developed by a partnership of the Offshore Renewable Energy Catapult, Scotland's Energy Skills Partnership, Heriot Watt University and Middlesbrough-based digital specialists Animmersion VR, the innovative system replicates in an entirely protected environment the working conditions experienced by wind turbine engineers operating on offshore wind farms.

Wearing immersive headsets, students experience the simulations of the real-life wind and weather conditions (as well as the potential drop!) they would experience at the top of a 7 megawatts offshore wind turbine. They see their own hands, feet and any essential training manuals as they attempt to undertake essential tasks. Unsurprisingly, offshore wind maintenance engineers can face extremely challenging conditions as they try to find, diagnose and ultimately repair faults, so an onshore classroom that enables them to develop their technical skills while beginning to understand and develop resilience to the additional challenges of the real-life experience, is obviously beneficial. Additionally, the training is significantly more cost-effective than if it were to take place in situ offshore.

As we explored in much greater detail in Chapter 5, tech-enabled wearables are beginning to make a big impact in the cruise industry. This also has an enhanced safety implication, for example with MSC Cruises' wearable bracelet. Relaxing parents are able to monitor their children's whereabouts in real time, as the sensors attached to their children's wrists interact seamlessly with the thousands of sensors in the ship's public areas. This means parents know exactly where their younger family members are at all times, so they can be assured that they are staying in the areas that they have been told to.

Within most Blue Economy sectors, reducing the risk of vessel collision is hugely desired. While it has never quite materialized, more money has been invested in technology solutions of late. Guidance Marine (part of the Wärtsilä family of companies) has developed the RangeGuard system, which aims to reduce the risk of collision by using radar to automatically detect objects within 300 metres of the vessel. RangeGuard sensors dotted around the vessel in effect act like a parking sensor or electronic bumper, providing warnings of potential hazards to crew within a user-defined range.

No safety net

Some approaches being developed and trialled in the fishing industry help to underline how problematic it can be to attempt to introduce game-changing technology within Blue Economy sectors. And if any sector needs such solutions, it is the fishing industry. The Fishermen's Mission, a UK national charity of over 130 years' standing, states that fishermen are some 115 times more likely to suffer a fatal accident than the rest of the general workforce, and reports that an average of 15 UK fishermen are killed or seriously injured every year.

Complementing these statistics, research released by the UK's Maritime and Coastguard Agency in July 2018 revealed that, in 2017/18, the fishing industry experienced a rate of 62 fatalities per 100,000 workers, meaning that commercial fishing is the UK's 'most fatal' profession. To hammer home the risk, the second most dangerous occupation (waste and recycling) had a rate of just 10.26 per 100,000 people, so in going to work fishermen face a death risk nearly six times greater than even the next most dangerous occupation. Grim reading indeed for the approximately 11,700 fishermen active in the UK, though it must be noted that this is a much smaller workforce number than the other comparison sectors.

Figures from the Royal National Lifeboat Institution demonstrate that, between 2010 and 2013, some 59 per cent of commercial fishing fatalities were due to a loss of vessel stability leading to capsize, leaking or swamping, with 30 per cent of these deaths occurring in the month of January when seas can be rough and water temperatures are at their lowest.

Globally, the Food and Agriculture Organization claims fishing at sea is probably the most dangerous occupation in the world, with over 32,000 fishermen perishing at work every year.

The Indian Government is investing significant resource in safety technology in the fishing sector, with some of the more notable activity being coordinated by the Indian Space Research Organization. However, one of their most prominent initiatives has, controversially, come up against the age-old barrier to safety and security at sea – user adoption. As, sadly, so often happens regarding safety initiatives, the pace of India's fishing technology programme has been driven by tragedy. On 30 November 2017, over 200 deep sea fishermen lost their lives in tropical Cyclone Ockhi. The devastating cyclone, which pummelled the coast with winds of up to 130 kilometres per hour, emerged in the Bay of Bengal and swept over Sri Lanka and south India, catching hundreds of fishermen unawares, many of them in rudimentary boats.

The geographical focal point of the industry's anguish was Tamil Nadu, the home of the vast majority of those lost. As the tragedy played out in the press and social media in real time, it became apparent that many lives might have been saved if it had been possible to send a warning to those already fishing far out to sea, to let them know that they were in grave danger. Fishing communities protested vigorously and, in the face of this pressure, the Government committed to invest in cutting edge transponders, developed by ISRO, to make possible emergency communication to minimize the potential for such a disaster happening again. A terminal and antenna are attached to the fishing vessel, and the system – which enables two-way messaging (with voice-based communication in development) in four languages – is operated via a smartphone app.

Crucially, in light of Ockhi, the system is satellite-enabled, aligned to NAVIC, the Indian Regional Navigation Satellite System, which covers India and a region extending 1,500 kilometres around it. By linking the system to the seven satellites in the NAVIC constellation, officials can not only monitor vessel location, but can also send targeted warning messages and receive SOS alerts from fishermen in particularly vulnerable areas, hundreds of miles offshore. As fishermen often go to sea in small fleets, one transponder can often service up ten vessels, if they are connected by their own additional internal wireless systems.

Seventy deep sea fishing boats in Tamil Nadu's Kanyakumari district were fitted out with the technology at the start of the fishing season in August 2018, following a smaller user trial on a handful of boats at the start of the year, in partnership with the Fisheries Department and the State Disaster Management Authority. While full details of those trials have not yet been made public, in November 2018 the Kerala State Government approved a budget of nearly £2.8 million to fund the roll-out of the system on 15,000 fishing vessels. By this point, a feature had been added to provide information to fishermen about international boundaries to ensure that they don't stray into other countries' waters and therefore begin to fish illegally, whether they intend to or not. In addition, a budget to provide 1,000 deep sea fishermen with satellite phones was also approved.

This technology was introduced during the peak of the Hurricane Okchi protests; Kerala's Fisheries Minister J. Mercy Kutty Amma inserted a note of caution when she stated that her department had provided fishermen with satellite-enabled safety technology previously, in the form of emergency search and rescue beacons that tapped into low Earth observation satellites to broadcast their location to emergency services when activated. However, the fishermen didn't actually take the devices with them, preferring to rely on 'natural warnings' instead. It was eventually discovered that these devices were left at home as playthings for the fishermen's children. Not only did this mean that they were not protecting themselves when out at sea, but this situation had a secondary drawback, as the trigger-happy children playing with the devices created a number of false alarms that wasted the time of the navy and coastguard.

This is no different, of course, to the distribution of earlier, non-digital technologies. The archive of news stories recording fatalities in the fishing industry is littered with examples of lives that could so easily have been saved if simple technologies such as life jackets or locator beacons had been utilized. In some cases they were not provided in the first place by short-sighted employers, but in other – even more galling – cases they simply were not worn even when they were available.

This is not a new problem. As the Fishermen's Mission point out, fishermen have had access to slightly more traditional equipment such as inflatable oilskins and lifejackets for a long time but, as they were considered too cumbersome to wear, they were simply discarded. A report in July 2018 from maritime welfare charity Seafarers UK found that simple safety procedures were regularly being ignored within ports, with less than 20 per cent of respondents to their safety survey admitting that they didn't wear a personal flotation device on a regular basis.

Forewarned is forearmed

Of course, the less time a fisherman is at sea, the less risk he will be exposed to, which is why emerging research by India's CSIR-National Institute of Oceanography is of such great interest. Researchers there are using satellite imagery and data collected from underwater equipment to push towards longer-term predictions of fish abundance.

As well as relying on its own instincts, the fishing community in India is also currently provided with daily fishing advisory notices from the Indian National Centre for Ocean Information Services. They use this data to decide where to go and fish; as it is only a daily service fisherfolk are obviously limited as to how far in advance they can plan their activity.

The new research programme hopes to use chemical and satellite data to push that eventually towards correct forecasts of locations of shoals of fish in sea waters nearly a month in advance. While the primary stated aim of this is to help protect fish stocks, providing fisherfolk with greater advance information can also help them to plan when and where they apply their fishing effort, enabling them to be more targeted (needing less time at sea to bring home the same amount of catch) and, where necessary, not feel the need to go out in rougher weather, if they feel more assured that they will be able to meet their quotas over the period of a week or a month.

Caught in the net

Safety is about much more than simply avoiding fatalities or serious injury, and this again is evident within the fishing industry, which continues to grapple with a sometimes unseen slavery issue. There are many factors that make fishing susceptible to criminality. The industry attracts a lot of vulnerable migrant labour, it by necessity straddles many international boundaries, and – with the oceans so large – many criminals consider it relatively easy (as opposed to land-based industries) to avoid detection, much less enforcement.

Against this backdrop, fishermen are at risk of forced labour, either by being trafficked or through wages and passports being withheld once they have joined a crew. With unequal power dynamics established, vulnerable enforced fishermen are subsequently more susceptible to harm – extremely harsh working hours and conditions, as well as psychological and sexual abuse – that can continue for months or years at a time. Several incidents have been reported of fishermen taking their own lives by jumping overboard rather than carry on enduring such abuse.

The problem has received significant profile in Thailand in recent years, which has subsequently been exploring what role technology may play in reducing the potential for slavery in its domestic fishing industry. United Nations statistics state that more than 50 per cent of Thailand's estimated 600,000 industry workers are migrants, often from poor adjacent countries such as Cambodia, Myanmar and Laos (Rose, 2018). According to the award-winning, Bangkok-based Labour Rights Promotion Network Foundation, as many as one in 10 commercial fishermen in the region are slave labourers (Rose, 2018), and Human Rights Watch released a damning report in January 2018 that claimed that human rights abuses, including forced labour, were still 'widespread' in the country's fishing industry (Human Rights Watch, 2018).

Simple technology solutions that the country has taken include legally mandating trawler owners to provide adequate communication facilities to enable crew to contact their families while on board, and introducing electronic bank transfer systems for wage payments, to verify that crew are actually being paid for their labour.

More innovatively, the Thai Labour Ministry announced in February 2018 that it had begun to use optical scanning technology as part of a much broader initiative to drive worker registration within the fishing industry. The irises of some 70,000 people who work on fishing boats had at the time been scanned in the previous three-month period, and further exploration of facial and fingerprint scanning had also begun. Such registration enables the authorities to ensure that fishermen are on the vessel they are registered to, and have not been 'sold on' to another vessel.

While there is still some way to go, such technology-enabled endeavours by the Thai Government have been well received in the international community, resulting – in part, and alongside many other initiatives – in the European Union withdrawing its 'yellow card' threat to ban Thai fishing imports. When you consider that Thailand's fisheries exports to the European Union are valued at well over three-quarters of a billion pounds, the investment in new technologies to help counter slavery can certainly be viewed as money well spent.

Cruise control?

Moving on from the fishing industry, there have been 96 recorded incidents of a person going overboard on a commercial cruise ship since 2000, according to statistics at Cruiseserver.net, with 17 per cent of those subsequently being recovered safely. It is worth noting that (according to analysis of the related story archive at *Cruise News*), passengers are most likely to fall overboard on the last night of their cruise, and the majority of people who fall overboard have been noted to either be intoxicated or ignoring clear safety warnings (and basic common sense) by undertaking patently risky behaviours such as climbing on railings.

Overboard incidents on cruise ships, then, average out at just over five incidents a year. While on balance this is a tiny number of incidents when placed in the context of the millions of passengers who take a cruise every year, it is very much a live issue for the industry.

The *perception* of safety within the cruise sector is obviously an important factor for related companies, as news and statistics for persons who fall overboard while on cruise ships (and associated fatalities) are always featured heavily in the press. As the more safety-conscious traveller may seek out such statistics when making purchasing decisions, there is also a strong commercial driver for cruise companies to do as much as possible to minimize personal risk to their customers, staff and contractors, over and above their inherent core commitment to safety.

However, the cruise industry has been subject to concerted criticism in recent years that it has not adopted more advanced emerging technologies coming to market. In this tussle is contained one of the key themes related to technology in the Blue Economy – when emerging technology is ready to scale across an industry.

The issue is perhaps best explored through the prism of a recent court case involving a death at sea. In January 2019, the US District Court for the Southern District of Miami ruled that Carnival Cruises were not responsible for the death of 33-year-old Samantha Borberg. The mother of four was said to be drunk when she went overboard at just after 2am, and her body was never recovered. While the acrimonious court case focused on whether Carnival were liable, as they continued to serve her alcohol when she was already intoxicated, another major issue was that the crew did not realize she had fallen into the water for 15 hours. Could available technology have made such a delay less possible?

Cruise companies have traditional utilized CCTV technology to assist in man overboard (MOB) incidents, but safety advocates thought that a significant breakthrough had occurred in 2010 when a federal law was introduced that required ships to install technology that automatically detects when a passenger or member of the crew falls in the water. The most important clause in the law, however, is that the cruise companies are mandated to do so 'to the extent that such technology is available', which has left everything open to debate. Most cruise companies in the ensuing years have simply refused to install the systems, saying that they are not market ready because they are either faulty, cumbersome or – more controversially – too expensive.

A MOB detection system that is considered 'complete' removes the need for a human being to witness the incident in real time, either in person or via CCTV images, by installing a network of sensors (including radar, infrared and/or video) designed to detect when someone falls overboard. Once triggered, the system automatically sends an alert to crew members, who can quickly review footage in the trigger area to ascertain if action is needed or if it is a false alarm, perhaps instigated by a bird or even a splashing wave.

Other tracking devices have been available on the broader safety at sea market for some time. For example, UK company SeaSafe Systems produces the Sea-Marshall AU9, an emergency marine location product that sends an emergency alert on 121.5 MHz, tracked by a base unit. The ship's crew use this homing signal to find and recover the casualty.

Similarly, the McMurdo Smartfind S10 is an AIS MOB device manufactured by McMurdo Marine that uses its inbuilt high-precision GPS receiver to transmit position information and a unique identity number back to the home vessel's onboard plotter. This signal will transmit continuously, complemented by a flashing LED indicator light to aid detection. However, the system requires the user to activate it by twisting its cap, so it would perhaps not benefit intoxicated cruise passengers.

However, these and similar products are limited when considering the utility needed for cruise passengers. First of all, as they are fashioned from bright orange plastic; they are not exactly the ideal cruise accessory, and are far from discreet. If, as we have seen, fishermen plotting dangerous seas do not want to be encumbered by such devices, holidaymakers in bikinis or at night, in their best holiday outfits, are very unlikely to wear them.

Second, they have no voice communication features, so onboard crew or the emergency services do not have the ability to confirm the requirement of a rescue operation. This means that an alert is not only potentially a huge waste of time for a cruise ship, but could be a risk to the emergency services as each time they send out a response rescue boat they put the crew's lives at risk and each launch costs a great deal of money.

However, more bespoke services are available. Specialist provider The MARSS Group – based in Monaco and London – has developed the MOBtronic device specifically for the cruise industry. The system relies on multiple radar sensor installations to monitor the ship wall and detect any falling object. To reduce the potential of a false alarm, analysis of each falling object takes into account variables such as speed and direction of the fall, and size and shape of the object in question. The company says that a crew alert is generated in 300 milliseconds. Any falling object is then tracked in the water while crew – who have been automatically alerted – can access the user console to review video footage and assess if a full MOB alert is required.

Even so, while making headway in other sectors such as luxury yachts and within navies, at the time of writing the MOBtronic has only been installed on one cruise ship. The industry still says that all available market offers are still not ready. Speaking in December 2018, industry body the Cruise Lines International Association confirmed – while committed to exploring technology options – their view remained that 'few systems have shown practical application on a cruise ship sailing the high seas'. Individual cruise lines have also commented that 'the promise is often better than the actuality', a phrase we often hear from potential Blue Economy customers in all sectors who, they say, need more hard evidence than can be provided by lab reports and found within sales brochures or PowerPoint presentations.

But when does promise become hard evidence? The MOBtronic technology – which won the 'Best Safety Product of the Year' at the IHS Markit Safety at Sea Awards in 2018 – evolved from a European Union research programme that supported over 7,000 test jumps, and extensive sea trials in a variety of marine environments over a five-year period

The company asserts that its technology has been proven in 45,000 hours of operation on cruise ships, that the system boasts detection probability of over 95 per cent and that it reduced false alarm rates to less than 0.3 per day.

Proof positive

Stepping aside from the specific of this case and this industry, this struggle underlines one of the key issues facing companies developing innovative technology – how to convince customers that an entirely new product is ready for market, fit for purpose and worthy of investment.

This can be particularly challenging when emerging technologies are so innovative that they are in effect creating a new market, or – perhaps more accurately – learning how best to articulate how they fit into and complement offers within existing markets. Autonomous systems may be seen as one such case in point. Start-ups that focus on what their shiny new technology can do (eg the number of days an unmanned vessel can remain at sea), rather than working on the (often very traditional) business or environmental problems they can solve (eg undertaking environmental surveys in extreme environments so no in-situ staff are required), will struggle to gain market recognition.

This factor should not be underestimated, as we see it time and again across the Blue Economy innovation space (though it is very much not an issue solely owned by emerging ocean technologies; this plays out across the entire innovation space). This pressure point creates, in effect, a double time lag for companies looking to bring a new product to market. First of all, once tech development is complete, it can take further time to arrive at the most sensible iteration of the product offer, especially as the business seeks user feedback to allow it to sculpt and reiterate the product narrative. That often takes more time than originally envisaged, so funds dwindle.

Second, as we have seen in the cruise industry, even when a strong business case is made, there is still often a time lag between the potential of the offer being developed and the rest of the world – the intended customer base – coming up to speed with the new technology and trusting it as a sound and sensible route to explore. It is at that point – when a concept has been developed but before a customer base has been matured to allow the product to be self-sustaining – that has been attracting closer attention from funders, to hopefully reduce the number of innovative concepts that add to the graveyard of promising ideas that did not make it.

In purely commercial sectors, it is often simply left to market forces to sort the good from the bad. However, an added layer of complexity is present in areas where there is also social or environmental gain to be derived from the use of new technologies, and this has drawn increased interest in recent years from a growing network of global funders.

Funding the future

Many 'tech-for-good' start-ups – not just in the Blue Economy – seek and receive pockets of funding in the beginning: £50,000 here and there for winning a competition celebrating new ideas within a specific sector, or more targeted funding to support companies to undertake a detailed feasibility study or to develop the concept or specific elements of the underpinning technology.

These funds come from a range of sources. Governments and related quasi-governmental agencies are important players in this space, looking to drive business growth as well as impact on specific social issues. Charitable foundations invest heavily in their long-standing areas of interest, and many corporations (especially within related tech industries) are also keen to be seen to engage and support tech-for-good initiatives, often channelling resource through their associated corporate foundations. While some focus on specific issues (eg the Gordon and Betty Moore Foundation focuses on environmental conservation, amongst other themes), or the application of specific technologies (the Vodafone Foundation invests up to £40 million a year in mobile-for-good innovations), other organizations such as the National Endowment for Science, Technology and the Arts (NESTA) and the Social Tech Trust are more interested in (in NESTA's words) broader 'innovation for the common good', most often driven by technology.

Invest to save

Within this funding landscape, there is a broader debate at play. For decades, NGOs in all sectors attempting to bring about positive social or environmental change – from promoting positive health programmes and encouraging literacy to fostering democratic engagement in emerging states or more sustainable practices in ocean economies – have been supported by charitable donations. Often this has involved quite 'traditional' projects where specialist staff are employed to deliver agreed programmes of activity over a set period of time and attempt to measure the impact they have brought about.

While not in any way attempting to downplay the role that innovation can play in such 'traditional' programmes – which always work hard to keep pace with changing demographic trends and ways of thinking, and often do so in very imaginative ways – the focus in some areas has been shifting to the longer-term potential enabled by embedded technology.

Why pay for an agricultural training programme to pass on key tips in crop management to 100 farmers in Kenya, when you can support the development of an app that does the same, but which then has the potential to reach many thousands of beneficiaries in a world where mobile phones (and, increasingly, smartphones) are becoming the norm for accessing information? iCow, based in Kenya, is an agricultural information service that does just that (Icow.co.ke, 2019). It started out as an SMS information service for small-scale farmers, and has now progressed to a smartphone app, having been supported in its journey by funding organizations as diverse as USAID, Accenture and The Indigo Trust.

Why support a time-limited 'clean our streets' campaign when you can instead invest your corporate foundation dollars in an app that will allow citizens to use the web or their smartphones to alert responsible authorities of fly-tipping or other environmental crimes? SeeClickFix is an app that started in New Haven, Connecticut, USA, by a technologist who 'reported a graffiti issue, and the city just ignored it' (SeeClickFix.com, 2019). From that starting point of personal frustration, the app is now live in over 340 communities in the USA, has been downloaded by over a million citizens and facilitated over 1.1 million citizen requests for action in 2018.

What's particularly interesting about SeeClickFix is that some of its early support came from the Omidyar network, the investment vehicle established by eBay founder Pierre Omidyar and his wife, Pam, to encourage positive social impact and encourage people to 'make powerful contributions to their communities'. While the Omidyar Network does make grant donations, it invested in SeeClickFix, in return for an undisclosed equity stake.

Such impact investment opportunities are also beginning to emerge in the Blue Economy. Companies investing in this way include Credit Suisse, Partnerships and Environmental Management for the Seas of East Asia, the Meloy Fund for Sustainable Community Fisheries (a division of the NGO Rare Group) and the Althelia Funds, which has a specific Sustainable Ocean Fund.

Such funds are very much to be welcomed. Not only do they provide the finance to drive emerging technologies towards sustainability, they also have a greater feel for some of the unique challenges relative to Blue Economy, and how companies need to set themselves up to weather associated storms.

Conclusion

Personal and professional safety of individuals at sea is a centuries-old problem. Despite many safety regulation improvements in recent decades, there is still no complete solution and deaths from drowning at sea and on other waterways are still too regular.

Companies developing innovative products for this market face multiple challenges. Developing products that work in harsh environments, at a cost that makes them viable when satellite connectivity can be expensive, is challenging enough. The next barrier is convincing potential buyers of the value of the product, which can take time. Even when that barrier is negotiated, navigating the final human element – of end users not bothering to use the products that have been bought for them – is again an age-old issue.

One potential solution to this perhaps involves looking more broadly at the utility of personal safety devices at sea. Any technology that only delivers safety features will always be at the mercy of the end user underestimating the risk that they face, or simply slipping out of the habit of following safety procedures.

Rather than developing mobile technologies that only focus on safety and subsequently get left at home for children to play with, how can such technologies be embedded in other systems that offer greater value to the end user – allowing them to communicate more freely, help them in other ways in their work or save them money? That will help to drive adoption of safety systems that will hopefully never need to be used but that still need to be carried. Similarly, the cruise industry may be able to take a leap forward when safety locators become miniaturized enough to be included in other wearable devices that passengers value and use more, as featured in Chapter 5.

Considering how to help the emerging Blue Economy tech sector to meet broader challenges, the role of development funding has never been more important. Especially, but not solely, related to the environmental domain, for decades funding has been channelled through NGOs in various parts of the world to deliver positive progress over a set period of time, but that paradigm is changing with the advent of more accessible technologies.

While there is a strong argument to be made that cutting edge technologies have the potential to provide longer-lasting benefits once the technology scales, technology on its own is not a silver bullet, and will also require ongoing costs to function and continue to deliver value once any periods of grant funding have finished.

This is not, then, just about investing in technology that can deliver compelling positive social or environmental impact at scale but finding models that can do so in ways that don't require continued ongoing funding and can progress to stand on their own two feet as viable, sustainable businesses.

That shifts the narrative from developing technologies that simply provide positive impact to doing so in a way that aligns with a strong articulation of business value to customers. Once these customers are convinced enough of the value of a product or service to start paying for it, a financially sustainable future for the underpinning business begins to look possible, ideally supported by a strong business plan.

Entrepreneurial founders of technology companies that have a strong social or environmental focus have many starting points. Some may start their journey based on their technical or engineering excellence. Others may have deep knowledge of and passion for the associated domains in which the technology operates, and the problem they are trying to solve with their technology. But that does not necessarily mean that they have the associated business strategy, business development and marketing skills to enable them to take their product to market. Often, in fact, these are skills that are not found within start-ups, and coming to that point of realization can make or break a company. This stage can often be the 'valley of death' for technologists: when they have a fully functioning technology but their money starts to run out before they have been able to convince customers to start to pay for their product or service so that they can take advantage of the amazing benefits it makes possible.

It has been interesting to watch the innovation funding space in the past few years, and to see how funders are learning these lessons. Most tech development growth funds follow a similar pattern. They are not only interested in testing understanding of the problem to be solved and how applicants plan to do so, but also how they are going to develop their company into a sustainable going concern than will not have to rely on funding in the longer term.

Whereas many NGO funding applications would often include as standard a question on sustainability (eg how the benefit of the programme would be continued post-grant), focus on this area has become much more acute in the tech-for-good space, and especially when looking to invest significant funds (eg anything over a million pounds) in technologies that promise to scale. Companies seeking development funds are now expected to display a strong understanding of the markets they seek to address, and how those markets function in practice. They have to produce sculpted definitions of business value that will underpin their sales journey, and – crucially – capture that journey within a business plan driven by a balanced budget.

The strategic shift of funders towards technology sustainability is useful not only to provide themselves with additional comfort that their investments are well made, but also to highlight the extra skills and experience required within innovative Blue Economy technology companies looking to scale.

References

Bansal, A, Garg, C, Pakhare, A and Gupta, S (2018) Selfies: A boon or bane? *Journal of Family Medicine and Primary Care*, 7(4), pp 828–831. www.ncbi.nlm.nih.gov/pmc/articles/PMC6131996/ (archived at https://perma.cc/F2EG-E9EV)

Human Rights Watch (2018) Thailand: Forced labor, trafficking persist in fishing fleets. www.hrw.org/news/2018/01/23/thailand-forced-labor-trafficking-persist-fishing-fleets (archived at https://perma.cc/CC6G-LTQK)

Icow.co.ke (2019) Home page. http://icow.co.ke/ (archived at https://perma.cc/LK95-YGM6)

Rose, J (2018) Thailand's slave fishermen: What's needed to solve the crisis? www.aljazeera.com/indepth/features/thailand-slave-fishermen-needed-solve-crisis-180911223139627.html (archived at https://perma.cc/QD5C-EBUE)

SAFETY4SEA (2018) Seafarer killed during container lashing in Port of Dublin. https://safety4sea.com/seafarer-killed-during-container-lashing-in-port-of-dublin/ (archived at https://perma.cc/5MV4-BGSU)

SeeClickFix.com (2019) Home page. https://seeclickfix.com/ (archived at https://perma.cc/88U8-C5DA)

13

Conclusion

Successfully navigating a sea of opportunity

We wrote this book to share our passion for and insights on innovation in the Blue Economy, rather than with a specific endpoint in mind. We knew which sections of the market we wanted to explore (as articulated in the definition of the chapters) and we had an already bulging database of technology innovations to explore (further extended greatly in the course of new research), but we were not seeking to prove a hypothesis. Rather, we wished to be led by the real-world examples of technologies showing promise within academia and pushing (slowly or at pace) into commercialization, bolstered by our ongoing discussions across sectors with technology innovators, policy formers, scientists, funders, operators and – as often as possible – those who rely on the seas and oceans for their livelihoods. All of which, of course, needed to relate back to the major needs and opportunities within the Blue Economy, which are becoming understood in greater detail year on year.

So what have we learned? Aside from developments within each market segment, which cross-cutting technologies are showing greatest promise across the piece? What consistent barriers are being faced? Are there opportunity themes that obviously present themselves, and what might unlock greater activity and progress? Many strong themes have emerged across these key questions.

The keys to successful product development in the Blue Economy

We are genuinely in awe of the technical know-how, creativity and tenacity inherent within so many innovators within the broad range of Blue Economy sectors with whom we generally engage while undertaking our various day-to-day activities and with whom we have come into contact in the course of writing this book. In assessing the progress being made in various areas, however, and how successful technologies were breaking through, we identified a number of consistent barriers faced by entrepreneurs when trying to bring new products and services to life to serve Blue Economy needs.

Perhaps surprisingly when considering the levels of technological advancement being progressed across various sectors, it was very rarely the technology itself that caused the main barriers. While time-consuming and intricate, engineering processes are well established and understood, and even when major technological breakthroughs were being sought or promised, the associated tensions were considered manageable in ways that other strains weren't.

Most of the more serious barriers clustered around the perhaps more intangible frustrations associated with explaining the new product or service, of reassuring individual potential customers (and sometimes entire industries) that the new offer could be trusted and was worthy of immediate consideration as it would provide a sensible return on investment – in short, concrete commercialization of a promising idea.

Several key approaches that make success more likely were apparent – either in learning from successful companies or, conversely, by studying the mistakes of those who were struggling to break through. While the journey of any technology innovator is long and hard, and myriad challenges have to be faced, we have summarized the most prevalent approaches that underpin success into a 'top four'. Those themes are:

1 *Be wanted: Meet a need*
   Embedding the approach to technology within a deep understanding of the need to be served or the problem to be solved will provide the greatest platform for success. If deep experience of the industry

being served is not contained within the core development team, research and collaboration to understand the true nature of the market is essential.

2 *Be user-led: Learn and pivot*
   Testing the initial idea with those with front-line experience is essential to making sure that the need can be met in a way that is operationally effective. Innovators need to be lean, and potentially be ready to 'pivot' the product in ways its customers suggest are needed. As growth comes, finding ways to stay in touch with front-line feeling, needs and emerging uses will also be required within markets whose needs can change quickly due to changes in environmental circumstances or the policy framework.

3 *Be sustainable: Develop the business model*
   Identifying who will buy or pay for a product is often not as easy as it sounds. Finding gaps in existing infrastructure that the product might add value to, fit in with or complement may be the key to success, rather than assuming it will disrupt or displace current offers entirely.

4 *Be patient and resilient: Customer take-up might take time*
   As we see repeatedly, the world may take a little time to wake up to the possibilities of a product or service that is being developed to meet a Blue Economy need, no matter how good or potentially transformative it is, or is promised to be. It is therefore undoubtedly essential to build in time within business plans for customers to understand what is being proposed and find budget to commission the product or service being offered.

We explore each of these in a little more detail below.

Be wanted: Meet a need

Any product will live or die by the size of the market it is trying to serve, and with Blue Economy products and services (which often but not always have a strong social or environmental element to them) that often relates directly to the size and scale of the need that they are trying to serve or the problem they are trying to solve. So solving that problem has to be at the heart of any offer if it is to be commercialized successfully and if it is to develop into a long-term business offer.

While this may sound blindingly obvious, it is surprising how often it can be overlooked or forgotten in the heat and turbulence of new product design, especially if the focus leans too heavily towards technological possibilities rather than addressing the core needs of the customer. When cocooned in a workshop, engineers can unwittingly let the process focus on what the technology is *able* to do rather than what it *should* do to meet the most urgent need of the customer. This can lead to the 'bells and whistles' that some technologies are criticized for playing up in shiny presentations at conferences, as they do indeed bring new technology into play, but not always to the greatest effect or to meet the most urgent need.

The best way to hard-wire the right focus into technology obviously starts with a deep understanding of the problem being met. This can often come from hard-earned experience in the particular sector being served, so is innate in the innovator. Even then, though, wrong assumptions can creep in over time that need to be challenged with fresh thinking. Alternatively, many innovators actually start with the technology capability itself, sometimes with insufficient knowledge of the industry it is intended to serve.

This can happen for a number of reasons. Sometimes, emerging technologists – especially, we have witnessed, those in the earlier stages of their career – are drawn to Blue Economy sectors due to the personal resonance of the potential social or environmental benefits that new products or services can provide. Tech companies of varying sizes often have the strapline of 'changing' or 'saving the world' and many Blue Economy sectors put wind in those sails. Alternatively, owners or developers of technologies developed for other sectors can be drawn to the Blue Economy by the size of the market opportunity, and attempt to repurpose their product or service – which may already be providing great value to customers in other sectors – to new cohorts of marine or maritime customers.

In both cases, this represents shaky ground on which to build a successful business. Whatever the new technology aims or claims to do, it only really has a serious chance of achieving reasonable scale if developers know how it can deliver its promised benefit within often very complex value chains and operational infrastructure. A deep understanding of the need to be served, the problem to be solved and the operational landscape within which it is to be achieved is essential. If broad relevant industry experience is not immediately available within the core development team, a crucial ingredient is missing. In order to mitigate the risk of substantial expenditure being wasted, steps need to be taken to identify relevant experience to guide direction and to inform the way the service is built. That can involve detailed desk research to understand the market better, or in-depth and soundly structured feasibility studies that engage and collaborate with the market to build a better sense of the nuances of operational need and delivery frameworks. Time and again this area is under-resourced, with the result being that projects simply fail.

This is an ongoing process and, even when initiated successfully at the outset of a project or product design cycle, will soon lead into the next point, based on further detailed engagement.

Be user-led: Learn and pivot

Once the broad landscape for development has been ascertained in detail, various levels of product and service development will be required. Depending on the nature and technical difficulty, this can take from weeks to years to complete. The key danger here is staying wedded to one idea that may need to be tweaked to meet the needs of the customer.

In short, innovators need to find ways to ensure that customers can play an integral role in the ongoing product development process. Some innovators setting out on development processes have told us simply that they intend to 'go where the market takes us'. To make that possible, innovators need to devise ways to keep their ears to the ground and work with and build trust with the market at every point of the journey. Some of this will involve low-level engagement, but then could scale up to active involvement in product testing and regularly asking established communities in the market what they need or what revisions they would like to see. Knowing your customers means knowing their problems, which in turns means knowing how to fix them. And regular engagement will nearly always provide valuable insights.

This engagement, then, needs to stretch far beyond the initial design stage, to encompass all stages of product life. Sticking resolutely to one static idea of what will provide value is fraught with danger, and may lead to a product or service that is obsolete as soon as it is launched. That is not to say that doing achieving this level of flexibility is in any way easy. Listening to what consumers, customers or beneficiaries want and need adapting an emerging product to meet those requests requires a particularly agile approach – developing ideas fast and, with rapid user feedback, refining a product quickly.

Often, this may not even necessarily be about what a product does, but about how it is accessed or used in the most beneficial way within real operational environments, so that it can become an essential provider of value rather than an additional chore. Why did the Indian fishermen not take the very expensive communication tools that had been provided for them after a serious cyclone incident, but rather left them at home for their children to play with as toys? Why have so many seafarers lost their lives because they didn't take advantage of the potentially life-saving products and services that had been developed and provided for them? How many apps have been developed for seafarers that never get out of the relevant app store because they don't really understand the harsh conditions they would need to be accessed at sea? While many factors will be at play in all of these examples, proper user-centred design and product development undoubtedly minimizes the risk of such a lack of product adoption.

As we have seen throughout this book, the needs of customers and entire industries can also change relatively quickly due to changes in environmental circumstances (eg the ever-emerging challenges of climate change) or the policy framework (eg the introduction of bycatch bans or regulations related to harmful emissions).

Information to drive product development can come from a variety of sources. Whatever technology is being developed, if it is possible to collect data through it, this should be collected safely from the start and analysed for insights wherever possible. This can be data automatically collected by the technology platform or service, or could be data collected through specific user evaluations. No matter what, any information that can be collected may well be worth it. It's important to note that such data collection need not be burdensome. With the right tools, many of which are freely available, data can be collected and analysed to add tremendous value.

Innovators, then, need to be lean, and potentially ready to 'pivot' the product in ways its customers suggest are needed, to cling to the highest levels of potential operational effectiveness as they move towards launch and beyond. As one environmental technologist, who had engaged properly with their intended customer base, told us: 'We did a couple of things at the very beginning that we thought were very smart but that turned out not to be very smart at all!'

Be sustainable: Develop the business model

Grounding a product or service concretely in evidenced market need, and subsequently adopting a lean methodology to develop the technology in line with the realities of operational use provide a healthy core platform for a sustainable business. The other major pillar to get right involves an entirely different set of skills, though. Developing a strong business around a product or service is essential if it is to achieve long-term sustainability, and the nuances of the associated tasks are again far from simple. Even getting the pricing right is a very difficult task – understanding the full costs of productization and, in the case of services, operational delivery is an ever-shifting process that needs to be undertaken at the same time as developing understanding of the costs that the market may bear.

To take the ocean survey market as one example, early talk from developers of autonomous and remotely operated vessels occasionally strayed into talk of these new platforms being truly 'disruptive' – that the owners and operators of existing survey vessels should be looking over their shoulders as these new services were cheaper, easier to deploy and put fewer staff in potentially dangerous conditions. The reality, as we are seeing, is that they are actually far more likely to become embedded as a complementary service. Survey companies tasked with collecting oceanography data can, for example, put autonomous vessels to work in a specific area before a full, manned survey, to help pinpoint the most useful areas to conduct the subsequent, more detailed research activities. Conversely, they could leave an unmanned vessel behind in that area when it appears that more data collection would be useful, but the main mission has reached the finite limits of its time at sea as the vessel is booked in for another job or its support staff are reaching the end of their allotted time. Traditional survey companies, then – originally framed by some as the fearful target of disruption – may actually more reasonably represent quite a healthy customer segment.

Even when the product or service's real, tangible benefit is really well articulated, when it is priced sensitively and when it is understood where it fits within the operational value chain, it may not be as simple as first imagined to actually get someone to pay for the service. Beneficiaries, users and customers are all often different, and it can take innovators some time to understand the difference and know where best to place their sales and marketing energies to unlock flows of revenue. The sooner it is recognized that 'value' is defined by the end user (who isn't always the buyer…) rather than the supplier, the more likely the innovation will be shaped to meet real needs.

There are several good tools out there to help innovators think through how they need to balance the various elements of their technology, market and business progress. One of the best-known is the Business Model Canvas structure that neatly and succinctly drives thinking about how core value propositions relate to identified customer segments, and how underpinning structure and processes (key partners, key activities, key resources, channels, cost structure and revenue streams) support commercialization.

Even using such good tools, however, may provide useful structure but that still does not make it easy. Considering who will buy or pay for a product is often not as simple as it sounds. If we take the example of Small Island Developing States as an example, they rely on the seas and oceans more than most but, in many cases, do not have large and available budgets to trial or purchase innovative solutions that may help them with their Blue Economy needs. For example, if the essential waters of a Small Island Developing State are littered with shipwrecks from a Second World War battlefield, it is well known that such wrecks degrade over time. If they contain large amounts of unexploded ordnance, or large amounts of crude oil, these wrecks are in essence an environmental disaster waiting to happen. Even so, the governments of the local islands in question simply may not have the funds available to hire a suitably equipped survey ship to conduct searches for vessels whose location is not exactly known, never mind conduct the necessary repair or salvage operations. In these cases, the picture becomes even more complex, as the end beneficiary may need to be brought onside to say they need the service in question, while a third party funder is secured as the 'buyer' of the service.

Within cases such as this, and in many broader areas of the Blue Economy, there is an interesting and ongoing conversation between the worlds of scientific organizations, NGOs, social enterprises and for-profit businesses.

When there is a real environmental need, which may be explored and addressed in part by NGOs or social enterprises (who are established as a business but do not take a profit or who have an 'asset lock' so that profits are not taken out of the company), the potential role of for-profit businesses can be seen as troublesome – 'profit' can be seen as a dirty word, or for-profit companies can be perceived as less favourable when compared to NGOs or social enterprises aiming to achieve similar aims.

Indeed, many Blue Economy entrepreneurs at the very start of their journey attempting to develop a technological approach to meet an identified social or environmental need often have to consider if they should incorporate from the outset as a business or a non-profit. The debate has certainly shifted in recent years, with the growing movement of social entrepreneurialism driving much more talk about businesses needing to have a purpose, and that you should make money by doing good for people, rather than just to drive profit. In fact, perhaps the most important framing should be not just whether a company is for profit or non-profit, but whether at its heart it is for *purpose*.

Even if an organization is built around a strong environmentally oriented mission, just assuming that setting up as an NGO is the 'best' way to go needs to be challenged. If impact and purpose are hard-wired into the company and its products and services, the structure of having a driven, profit-seeking board and investors can be helpful in developing a very focused and well-structured entity that develops products and services that work hard to articulate their value so that they can be sustainable in the marketplace without the need for the kind of grant funding that may dry up at any time.

Many, in fact, adopt a hybrid approach to funding – seeking and securing grant or aid funding in the early stages (sometimes even setting up a non-profit arm to enable them to accept those funds), and then converting wholesale into a business that derives its finance from customers or investors, once they are ready.

All of these considerations are an essential part of an emerging businesses conversation between the business itself and the market as it tries to find its place in the world, so that it has a business model that will allow it thrive and deliver value where it needed.

Be patient and resilient: Customer take-up might take time

Like all industries, the Blue Economy is littered with what appear to be brilliant ideas that didn't quite make the grade and quietly die out before they can drive towards the kind of market adoption that make them viable as going concerns. Quite often, this can be when start-up companies enter what is commonly known as the 'valley of death'.

As we talk so regularly both to the entrepreneurs themselves and to the individuals, governments, companies and sometimes NGOs to whom they are trying to sell a new product or service, we are in a position to appreciate these frustrations in the round. Often, we hear frustrated tech developers talk about their potential customers having a 'lack of sophistication' in understanding the full value that their new and brilliant technology can provide – that they are not technologically minded enough to be able to grasp its potentially transformative effects. Of course, this more usually signals that the entrepreneur has not yet developed the right, compelling way of describing their product or service. More worryingly, it may be that they have articulated perfectly the kind of value that they can deliver, but it just isn't good enough to make a difference to a customer with a limited budget.

Other ways of describing this critical phase of a company's development include customers being seen as too conservative, risk averse, individually slow to decide, or considered to be mired in inefficient decision-making processes within a larger company framework. Many struggling companies (and, indeed, many who eventually sadly fold) prefer to see themselves as being 'a bit ahead of their time', tragically caught in the time-lag between a product being available and it being understood, desired in the market and commercially successful. An understanding of the likelihood of such a delay would ideally be factored into a realistic business plan.

While some of this can seem like sour grapes, for most of the innovators we speak to it is simply an essential part of the learning journey and the ongoing, critical conversation that they have with the market as they hone their commercial offering.

Looking at it from the perspective of the potential customer base, it becomes clearer why such a time lag may occur. As we have been told repeatedly, even customers with a tightly defined need can struggle under the weight of the number of different approaches it has to choose from. For example, a shipping company may have an urgent need to reduce the amount of emissions it releases into the atmosphere so that it can comply with new regulations that would otherwise restrict its ability to trade. Its management understand this need, make budget available, and empower appropriate staff to start procurement processes. As the many technologies being developed to meet such needs – especially when driven by the demands of new policy amendments – can take wildly differing approaches as to how to provide value, it understandably takes time to scour the market of available offers, decide which offer might be most appropriate to the specific set of circumstances, and negotiate suitable commercial arrangements.

Within the cracks of these issues, another problem often expressed to us is what has been defined as 'pilotitis' – where lots of development money or grant funding is suddenly made available for a particular type of technology or to address a specific problem within an industry. At that point, the customer base may be approached regularly by energetic start-ups with working prototypes that they wish to put to the test in real-life operating conditions.

From the buying industry's perspective, this is not helped by the natural wastage amongst such companies – not all of them will succeed, and many will disappear after early trials. While perfectly natural and logical within the innovation environment, this can happen in plain sight, further feeding into the purchasing manager's mind to tread carefully with young companies.

While all of the very real challenges we set out in this chapter have the potential to paint a picture of constant tension and uncertainty, that is very much not our experience of the Blue Economy technology industry. There is something about the excitement of new technological advancement and the potential environmental gains that may be made across Blue Economy sectors that allow the space to remain vibrant, energetic, positive and retain a sense of both balance and good humour. What is true, though, is that the time lag between the product being available and being purchased at scale is very much a fact of life, as potential customers may indeed take a little time to recognize the possibilities of the product in development, or may wisely wish to take their time over investment decisions. This therefore needs to be addressed in development plans and revenue projections.

Addressing the Blue Economy funding network

Finally, it is worth taking some time to address some of the broader infrastructure that supports the development of Blue Economy technology development and innovation. This is the funding network that understands and believes in the potentially transformative effect of new technology within the ocean environment, and looks to invest in it to maximize its growth.

Most emerging technology innovations in the Blue Economy – especially those that are being developed within start-up companies – receive pockets of grant funding in the early stages of their development to write a feasibility study, prove the concept, engage with the user community, for very specific elements of technology development or, sometimes, just for winning a competition based on the novelty or promise of the idea. While most companies are extremely grateful for this early financial support, many also complain about the administrative and reporting burden that can come with the award of such grants. In our view, though, this is often for very sensible reasons. All funding bodies will have made their own mistakes over the years, and it is perfectly reasonable for them to put in place structures to help them minimize the potential for their investment to be wasted. Perhaps, sometimes, early stage innovators are too fledgling to fully understand why such demands are made of them, but in our view funders well understand the pressures on innovators and are actually looking to push them to address some of the key barriers that we have set out earlier in this chapter.

Many funding processes focus in on similar themes when asking companies to apply. They usually ask applicants to articulate the size of the market they are looking to address; they find ways to look at the originality of the idea and how innovative it is; they assess levels of technology readiness and an understanding of the development steps still needed to be undertaken; they ask for an overview of the team behind the idea to make sure the necessary skills are in place to realize the potential; crucially, given failure rates, they usually ask for statements of financial health, quite often even asking to see copies of up-to-date bank statements, to ensure the applicant has enough working capital to live on both through the period of funding and beyond.

Forms aside, many major funders complement the written application process with face-to-face presentations so they can see for themselves the levels of confidence, articulation and leadership available within the development team seeking funds. This therefore brings into focus something that can't be easily be assessed on paper alone – personality.

As we have set out within this specific chapter, and throughout the book, the road to successful product development within the Blue Economy can be extremely challenging. Building on the oft-heard line that 'funders don't give to projects, they give to people', many funders like to be able to see for themselves the personal qualities of the founders or co-founders to whom they are being asked to provide their resource. Funders have told us that they see this as incredibly important; if they can see that a company leader can communicate and engage well, that they can tell the story of their product and its benefit in a compelling way, then they will be able to do so in other arenas, in front of other funders, perhaps commercial investors and, eventually, within the marketplace in front of customers or important industry stakeholders. While non-charismatic business owners can of course employ others to take on such external, public-facing leadership tasks, that skill certainly needs to be somewhere in the mix.

The focus and weighting of these questions vary in direct relation to the relative maturity of the idea or how far along the road to commercialization it is. Considering all the companies we have engaged with on these agendas, perhaps what matters most relates to what might be described as an understanding and a sense of ownership of the barriers that lie between them and full commercialization of the service. In broad terms, companies can either address the questions that are asked of them because they are being asked to by the funder (ie treating them as 'tick-box exercises') to ensure the appropriate funds are released; or they can properly embrace the challenges set out and work as hard as they can to address them, because they truly understand that these are questions they need to answer fully in order to become a successful business. To provide an example, that can mean the difference between producing a business plan as part of a funded programme that gives very rosy revenue projections because they think that level of optimism is what the funder wants to see, and really working hard to be honest and transparent about the true levels of investment needed to develop a product or service fully, build in lag between when they start addressing the market and when sales will come, etc. Doing the former may well unlock a grant payment, but it actually misses an opportunity to address an issue that will need to be addressed at some point.

Incubation and acceleration

Some funders are addressing this issue by establishing a much closer relationship with the companies they seek to support. We are always pleased to see the launch of new technology incubators within the Blue Economy. As we have set out in these pages, the range of skills and experience needed to master technological, business and specific domain challenges are significant. Neither technologies nor companies very often come fully formed. Tech incubators look to acknowledge this by doing more than just offering money, but rather by blending such investment with direct advice and support for companies looking to grow. This usually involves cross-cutting expertise being made available, from technical support to help developing business plans or financial projections, or facilitating engagement with industry partners or potential financial investors (and helping to prepare emerging companies to make the most of these opportunities). The European Space Agency operates a range of such Business Incubator Centres (ESA BICs) across it network, for companies looking to utilize space-based assets. We're please to say that many of these companies focus on developing products for the Blue Economy, but also benefit from the opportunity to engage with start-ups in other sectors to draw learning across a non-competitive environment.

While the ESA BIC network focuses on the technologies being used, there are also a growing number of domain-specific incubators or accelerators. One such, called HATCH, is adopting an interesting model. Focusing on the major market opportunities associated with the aquaculture sector, they invite applicants to pitch for a blend of financial investment and business improvement support. One interesting element about this accelerator is that they have decided to follow an itinerant model – not basing themselves perpetually in one place, but launching in several countries over time. Their first location was, understandably, Bergen in Norway, but they have since branched out to Singapore and Hawaii. The HATCH offer includes money (€50,000), but also incorporates mentoring, industry connections, product development, office space and domain expertise.

When launching the Bergen HATCH office, a spokesman for the company behind the initiative, Alimentos Ventures, made clear why they thought the hands-on accelerator model was required, rather than just an injection of hard cash alone (Cosgrove, 2019). He drew a direct comparison with the IT sector where, he said, entrepreneurs are far more in tune with what investors want and how to speak to them – they go to business school, they understand about investor readiness, but that level of business acumen wasn't evident in as strong a way within the aquaculture sector (as we see continually across many Blue Economy sectors). Hence HATCH has begun to round out innovators' technical skills and domain expertise with the broader business skills required to commercialize and tell the story of their technology and the impact it can have in a far more commercially compelling manner.

The key enabler: A shift in investment priorities?

If that is the challenge facing the innovators, there is probably one area of development we would welcome within the funding world. We are generally hugely supportive of the funding framework as it addresses the Blue Economy. While we engage with and support – as the content of this book underlines – activity across the breadth of several Blue Economy sectors, there are many consistent themes across the funding landscape. Funds are variably available from governments, from broader entities such as the European Union, from charitable foundations, from individual philanthropists and, in many cases, from philanthropic foundations aligned to companies or multinationals, as part of their corporate social responsibility commitments.

Their intended impacts vary. Many government funds look to see returns on their investment in terms of new jobs created in developing and commercializing new services; for these funds, then, the specific focus of the selected industry itself may be of less importance than its ability to commercially consume new products and services that create thriving businesses and therefore employment opportunities. More often, though, funders want to see a real difference made in their areas of expertise and interest, be that in reducing greenhouse gas emissions in shipping; doing more to protect the lives of seafarers; helping to eradicate illegal fishing; or stopping the unintended bycatch of endangered fish species in trawler nets. Such funds can be more demanding about understanding the markets being addressed and the problems to be solved, as they appreciate in more finite and comprehensive detail the nuances of the industries being addressed and the operational realities therein.

While accepting that there will always be natural 'wastage' in a competitive landscape, our engagement in the Blue Economy also suggests that there appears to be a pattern where many emerging businesses struggle for further development costs at a crucial time, revolving around the double time lag set out earlier, which affects the maturing companies more than the earlier stages of development. The first lag relates to the time it takes to arrive at the most sensible iteration of the product offer – especially as companies seek user feedback so that they can reiterate. That often takes more time than originally envisaged, so funds dwindle. Following hard upon the heels of that issue comes the time lag between the potential of the offer being developed and the rest of the world – the intended customer base – coming up to speed with the new technology and trusting it as a sound and sensible route to explore. It is this point – when a concept has been developed but before a customer base has been developed to allow the product to be self-sustaining – that may warrant closer attention from funders, to hopefully reduce the number of promising concepts that add to the graveyard of 'bleached bones' across the tech-for-good sector, and not just as it applies to the Blue Economy.

This may involve shifting the balance of funding slightly from earlier stage companies (who are quite well served with early stage innovation funds that support proof of market opportunities and feasibility studies) to those gusting towards the higher levels of technology readiness but still not yet ready to fully commercialize. The stakes have usually by necessity increased for such companies – they may have had to invest significantly in technology development, and in doing so have a team of skilled, committed staff that – while no doubt an excellent resource – become a 'beast to be fed' that creates all sorts of pressure while the double time lag kicks in.

It is not necessarily that these companies fail; many find their way after a period of readjustment, of slimming down the cost base, or of finding ways to collaborate or rethink their business models. While that may be a necessary evil to ensure they are fit for purpose, it often strikes us as a frustrating delay in bring new, potentially transformative ideas into play. What matters most is that there may be a smoother and more accelerated route to get new technologies out of the workshop and into operation. As we have seen throughout this book, the challenges – and opportunities – facing the Blue Economy are many, varied and increasing in breadth, complexity and importance. As we finish this book, a new wave of environmental activism is breaking out across the world, and looks set to continue in the months and years ahead. Anything more that can be done to help relieve pressure in one of the areas that stands to be buffered most by climate change – the seas and oceans and all that depend on them – is undoubtedly to be welcomed.

The greatest gifts

While we purposefully did not set out to favour any one set of technologies over any other, and followed what is happening in tech market sectors, it became obvious through the research that several clusters of technology are having a major positive effect across the Blue Economy, having core functionalities at their base but being put to use in many different domains. These include:

- autonomous and remotely operated systems that in particular promise reduced costs, greater ease of deployment and more comfort on health and safety risks, especially in harsh environments;

- Earth observation capabilities that capitalize on the greater availability and reduced costs of space-based assets and data to drive new levels of value in a wide variety of ways;
- Big data systems that use powerful new machine learning and analytics capabilities to draw in from a wide range of disparate sources and convert it to actionable intelligence.

Each of these also provides greater prominence to the need for successful marine and maritime connectivity, so that new technologies in development have the potential to offer force multiplier benefits in real- or near-real-time. Finally, a greater reliance on systems powered by data massively increases the need for need for cyber resilience in shipping, ports, autonomous vessels, satellite systems and all supporting technologies and related sectors.

If we were to bring all of these under one umbrella, we would undoubtedly say that situational awareness is king within the Blue Economy, and that its potential is being given a major boost with the continued emergence of new technologies. By this we mean that all these systems can combine to fuse data from new and traditional sources (including from space) to provide operators and scientists with the best possible and most up-to-date and information-rich context with which to make informed decisions. That could mean providing a national coastguard with a detailed picture of the recent activity of a ship entering its waters, made possible by machine learning correlating traditional port records with satellite positioning information. It could mean a ship's captain being provided with an algorithm-produced suggestion of how best to navigate his or her upcoming course, based on a blending together of many years of historical fuel efficiency and wave/current records aligned to the upcoming weather forecast. It could mean providing a ship's engineer with a sensor-driven alert that, based on an analysis of recent *in-situ* engine vibrations and the planned journey ahead, indicates that an urgent replacement of a part may be required in order to ward off failure and a period of down time as essential repairs are made.

All of these possibilities offer great promise to a broad, complex and in many ways booming industry. What is also clear is that this is also just the beginning. Each technology, within each company, within each sector, is beginning to collect more data than ever before. This is being accelerated both by the broad sweep of digitalization and also by the rapid emergence of analytical tools that drive the appetite for its collection, as it can now be turned into useful information. Once that sea of data becomes as large as the very oceans it is looking to serve, and further collaborations to explore it for mutual benefit emerge, the future will be very bright indeed.

Reference

Cosgrove, E (2018) Aquaculture start-up ecosystem takes a step forward with new accelerator. https://agfundernews.com/aquaculture-startup-ecosystem-accelerator.html (archived at https://perma.cc/42L8-L2FP)

INDEX

A.P. Moeller Maersk 8 *see also* cyber attacks *and* Maersk
Abu Dhabi Ports Company 32
   and use of drones in its ports 32 *see also* drones
Africa/African
   2014 ebola outbreak 191
   ban on plastics in Rwanda 155–56
   coastal nations 139
   and FISH-i Africa 192, 193
   river systems carrying plastic 149
air pollution 8, 36, 37–39, 41 *see also* United States (US)
Airbus 16–17
   and wind power 16–17
AISs (automatic identification systems) 91, 93, 94–95, 97–98, 187
   seen as regulatory nuisance 95
   tracks 188
   turned off by smugglers 97
   and valid reasons for turning off 95
Alimentos Ventures 254 *see also* incubators/accelerators
   and Bergen HATCh office 254
Amma, J Mercy Kutty (Fisheries Minister, Kerala) 225
Anderson, D 115
Anderson, M (Director, Dallas Museum of Modern Art) 79
Antwerp
   Echodrone trials in the Port of 34, 136
   and PortXL concept 43
aquaculture (and) xi, 2, 109–28
   Aquabyte 116
   communication breakdown 117–19
   fish cages 121–22 *see also* United States (US)
   and AquaStorm project (Mowi) 121–22
   'escape-proof' – the Aquatraz 121
   the global race (and) 112–15 *see also* Indonesia
   animal health, hygiene and water use 113–14
   sea-lice 114–15
   water conditions 114
   imprisoned in steel 121–22
   meeting world food needs in environmentally-sound manner 109–12 *see also* reports *and* World Bank
   potential negative effects of 123–24 *see also* articles/papers *and* Hjul, J
   protection of aquaculture cages/nets *and* net-cleaning 119–21 *see also* robots
   rise of harmful algal blooms (HABs) 115–16
AquaStorm project 121–22
ARA *San Juan* 197–202, 206, 207, 216–17
   search for 197–200
   search and recovery team 197
   wreckage located 200
Arctic Expedition Cruise Operations, Association of 76
articles/papers (on)
   benefits of a growing aquaculture sector (*BioScience*, 2018) 124
   locations holding nearly 70 per cent of total tidal potential in measured areas in Indonesia (IRENA, 2017) 52
   maritime surveillance and security issues (*National Security Journal*, Harvard Law School, 2015) 92 *see also* Salerno, B
   potential of wave power technology in Indonesian archipelago (*International Journal of Mechanical Engineering and Technology*, October 2017) 52
   tidal potential in IRENA's renewable energy roadmap for Indonesia (2017) 52
artificial intelligence (AI) 15–16, 22, 25, 48–49, 78, 106, 125, 165–66, 167 *see also* Google *and* Stena
   airborne 182
   used in shipping industry 26
   Zoe: virtual personal cruise assistant (MSC Cruises) 84–85

ASEAN (Association of Southeast Asian Nations) 50
Asia (and) 139, 149 *see also* Southeast Asia
    domination of world seaborne trade 7
Asia Pacific (and)
    plastic 152
    region 49, 152, 203, 214
    rise in energy consumption 49
    wrecks in South Asia Pacific 203
Asian Development Bank 96 *see also* Tonga to Fiji subsea cable
    and funding of sub-sea cable (jointly with World Bank) 211
Atlan Space drones (and) 182–83
    daily scanning of ocean 182–83
    deep learning model identification and analysis of vessels 183
Atlantic Telegraph and cable from Queen Victoria to James Buchanan (US President) 209
Atlas of Marine Protection 163
Attenborough, Sir D 154
    and *Blue Planet II* (BBC) 154
Australia (and) 121 *see also* Telstra
    collaboration in development of shield to block sun's rays to protect corals 166
    Great Barrier Reef Foundation 166
    Institute of Marine Science 166
    University of Melbourne 166
automatic docking 35
autonomous and remotely operated vessels 103–04 *see also* Rolls-Royce
    potential of 103
    safety, security and attractions of 103–04
autonomous underwater vehicles (AUVs) 62, 197, 198, 206–07
    HUGIN 207
    Seabed Constructor 197, 198, 199, 200

Bali (and)
    ban on single-use plastics 155
    tourists 54
Baltic and the International Maritime Council 28
Bansal, A 220
Barbuda 140, 142, 143, 219 *see also* Antigua
bathymetry 196, 201 *see also* satellite-derived (SDB)
Belgium: offshore wind market 55
Berger, M 163

Bergman, J 28
Berners-Lee, Sir T 189
Big Data 25, 87, 106
    potential of 189
bird populations, adverse effects of offshore and other wind farms on 65–66
blockchain technology/ies 25, 42–43, 189
blockchain-enabled technologies 182
the Blue Economy 1–5, 31 *see also* NLA International (NLAI)
    and Blue Growth 3
    embracing complexity 2–4
    global value of 2
    as an ocean of opportunity 1–2
    as reliant on digital infrastructure and data-driven services 8
    and threat from rising global temperature 164
Boffey, D 36
Bora, T 72
Bousso, R 39
BP *and* phishing emails 9 *see also* Mason, G
Branson, R 162
Braun, A 166
Briddon, M (Engineering Manager, James Fisher Mimic) 19–20
Buchanan, J (US President) 209
Burberg, S 229
    Carnival Cruises not responsible for death of 229
Burden, D (Director General, Naval Materiel, Argentine Navy) 195
Burgess, A 42, 148

Cabico, G 110
Canada (and) 121
    Canadian Hydrographic Service (Fishers and Oceans Canada) 135
    National Strategy to Address Abandoned and Wrecked Vessels 206
    Transport Canada Abandoned Boats Program (launched May 2017) 206
    Wrecked, Abandoned or Hazardous Vessels At (Bill C-64) 2017 206
Carbon Trust Offshore Wind Accelerator 60 *see also* studies
Caribbean islands as vulnerable in terms of disasters per capita/land area 139
Carlsson, L (Head of Artificial Intelligence, Stena) 14, 16
Carney, B 122

Carnival Corporation/Cruise Line/fleet
    (and) 80–84, 86–87
  BOLT™ 71
  digital and Big Data innovation
    experiment 87
  Ocean Compass app 81
  Ocean Medallion™ 84, 86
Caruana, C 164
catastrophe: Deepwater Horizon (2010)
    204
Cayman Lidar surveys 133
  and multi-beam echo-soudners
    (MBEs) 133
Chambers, S 7, 25 *see also* reports
chapter conclusions (for) *see also* conclusion
  aquaculture 122–25 *see also* articles/
    papers; reports *and* research
  the cruise industry 86–87
  hydrography and bathymetry 144–46
  maritime surveillance 106–08
  ocean conservation 166–69
    and crowdfunding 167
  offshore renewables 67–68
  ports and harbours 44–45
  safety of life at sea 235–38
  shipping 24–28
  sub-sea monitoring 215–17
  sustainable fisheries 192–93
Chile, breach of salmon farms in 116, 119,
    121
China (and) 16, 98, 111
  economic expansion in 7
  Ministry of Transport's aim to increase
    number of cruise passengers to 14
    million 73
  offshore wind market – third in global
    offshore rankings 55
climate change 3, 50–51, 54, 92, 93, 106,
    120, 124, 138, 162, 163, 165,
    180, 244, 256
coastal populations, dramatic increase in 139
Coffey, H 155
Cohen, P 218
Comprehensive Nuclear Test Ban Treaty
    Organization (CTBTO) 196,
    199–200
  Hydroacoustic stations: HA10 (Ascension
    Island) *and* HA04 (Crozer) 199
Compton, L 120
conclusion (and) 239–58
  addressing the Blue Economy funding
    network 250–53
  be patient and resilient: customer take-up
    might take time 248–50

  be sustainable: develop the business
    model 245–48
    and Business Model Canvas
      structure 246
  be user-led: learn and pivot 243–45
  be wanted: meet a need 241–43
  the greatest gifts 256–58
  incubation and acceleration 253–54
  key enabler: a shift in investment
    priorities? 254–56
  keys to successful product development
    in the Blue Economy 240–41
    1. be wanted: meet a need 240–41
    2. be user-led: learn and pivot 241
    3. be sustainable: develop the business
      model 241
    4. be patient and resilient: customer
      take-up might take time 241
  successfully navigating a sea of
    opportunity 239
condition monitoring systems 18–19
  ClassNK-NAPA GREEN 19
Coral Triangle Initiative (CTI) 205
  and 'Ring of Fire' region 205
Cosgrove, E 254
the cruise industry (and) 71–88
  back to the future 76–77
  coming in from the cold 75–76
  floating on the digital revolution 77–80
  getting personal 80–82
  greener and cleaner 74–75
  Okay, Zoe? 84–85
  pushing the digital boundaries of the
    customer experience 71–73
  reaching for the stars: connectivity,
    Wi-Fi; satellite options *and*
    VSAT 85–86
  safe and secure 82–84
Cuvier's beaked whale, cause of death of 147
cyber attacks (and) 8–9
  A P Moeller Maersk 8
  BW Group: breach of cyber defences
    8–9
  NotPetya ransomeware 8
  ransomware attack (Port of San Diego,
    2018) 9
cyber security 9, 12–13, 25, 27, 28, 34
  maritime sector-focused, KPMG and
    Kongsberg partnership on 12

Daley, J 162
Dallas Museum of Modern Art: free
    admissions and membership
    initiative 79–80, 83

262 INDEX

Daly, J (Irish Minister for Mental Health and Older People): call for 'selfie seats' 220
Dao, T 110
data and open data hubs (eg London Data Store) 189
data sharing and donations of data (and)
    between countries 192
    FISH-i Africa (funded by Pew Charitable Trusts) 192
    Global Fishing Watch, donation of data to 191–92
    Global Fishing Watch open data platform 191
    Leonardo diCaprio Foundation 191
    VMS data on 1200 small fishing vessels (made available by Peru, 2018) 191
data theft 8
definitions of the Blue Economy, variations in 1–2
Denmark (and)
    cyber strategy for shipping industry (Ministry of Industry, Business and Financial Affairs, 2019) 13
    Danish Maritime Cyber Security Unit 13
    offshore wind market 55
Di Caprio, L 162
Dietrich, T 159
The Digest of UK Energy Statistics (Dept. for Business, Energy and Industrial Strategy) 55
Digital Ship Chief Information Officer conference series 25
    and CIO Forum 27
disasters *see also* hurricanes
    Port au Prince earthquake (2010) *and* Digicell experiment 190–91
    West Africa Ebola outbreak (2014) 191
Dodds, L 44
Donald, A (CEO, Carnival) 82
D'Onfro, J 15
drones 32–33
    and drone-delivered imagery for US fleet 33
    technical dexterity of 32–33
    and terrorist groups 33
Dunant, H (founder of the Red Cross) 210

EASOS (Earth and Sea Observation System) 99–100
East Flores (and)
    prospective bridge in Larantuka Straits 54
    Tidal Bridge BV 54

Echodrone (autonomous survey boat): trials in the Port of Antwerp 34, 136
Economic Co-operation and Development, Organization for ()ECD) 1
economic trends (Community of European Shipyards: CESA, 2016) 3
Edge (Celebrity Cruises) 72 *see also* Yousafzai, M
Edwards, J 11
Einstein, A 77
European Marine Energy Centre 53, 59
European Space Agency (ESA) 118
    Business Incubator Centres (ESA BICs) 253–54
    guide to Earth observation 183
    NAVISP programme 11
European Union (and)
    aquaculture in EU markets 110
    ban on some single-use plastics 155
    definition of maritime surveillance 91–92
    funding from 254
    increase in governance pressure on addressing bycatch issues 175
    Interreg North West Europe programme 59
    potential fines on UK for failure to meet air quality targets 37
    research programme (MOBtronic technology) supporting over 7,000 test jumps 231
    withdrawing 'yellow card' threat to ban Thai fishing imports. 228
Exclusive Economic Zones (EIZs) xi, 97

fishing fatalities 223–26 *see also* safety of life at sea
    in Cyclone Ockhi (2017) 224
    due to loss of vessel stability 223
    leading to investment in transponders for emergency communication 224–25
Finferries 22 *see also* Rolls-Royce
Fishermen's Mission (UK) 223
    and discarding of potentially life-saving equipment 226
Fletcher, R 114, 118
Flettner, A 77
Food and Agriculture Organization (FAO)
    estimates of percentage of total global fishers catch discarded annually 173–74
Ford, H 162

Forum for Future Ocean Floor Mapping: opened by Prince Albert of Monaco (2016) 130
Fourth Industrial Revolution 189
  becoming more pervasive 187
Foxwell, D 20
France 55, 59, 155, 167, 205, 210

G-20 group of developed nations 50
Gabbatiss, J 152
Garcés, J 116
Gatward, I 112
Gaworecki, M 158
GEBCO (General Bathymetric Chart of the Oceans) 129–30, 131 *see also* Japan
  Seabed 2030 initiative 134, 146
geographical information system (GIS) 129, 137
  and specialists Esri 144
Germany (and) 55, 62, 156, 194
  effect on birds of offshore wind projects in German North Sea 65
  submarine *U-166* 204
Ghaddar, A 39
global energy consumption *and* demand 49
Global Fishing Watch 191–92, 193
Global Industry Analysis: on future growth of offshore wind capacity 56
Global Maritime Issues Monitor 27
global recession, challenges of 7
Global Wind Energy Council report on offshore wind markets 55
  and UK as the largest 55
Glumac, T 120
Google 15 *see also* Pichai, S
  and Dunant Cable (2019) *see also* Dunant, H
  Earth 144
  investment in 14 new subsea cable systems 212
  investment in planned Pacific Light Cable Network 212
  public comment on reasons for investments in cable systems 212–13
Godfrey, M 125
Gokkon, B 114
Gove, M 111
Great Ocean Cleanup 151, 153, 154, 157, 168
Great Pacific Gyre/Garbage Patch 150, 151
Griffin, A 36

growth of global hydrographic survey equipment market (Markets and Markets) 131–32
Guidance Marine RangeGuard system for collision risk reduction 222–23

Hapag-Lloyd cruise expeditions using marine gas oil (MGO) 75
harbours *see* ports and harbours
harmful algal blooms (HABs) 115, 118
Hawkins, L 35
Heaton, B 26
Hendriksz, V 78
HiLo *see* predictive modelling
Hjul, J 112, 123
Hoshaw, L 158
hurricanes
  and cyclone Ockhi 224
  Dominica (2017) 139
  Florence 136
  Irma 140, 141, 142,
  Jose 140
  Maria 139
hydrography and bathymetry (and) 129–46
  see also Cayman Lidar surveys *and* Lidar
  actionable intelligence *and* bathymetry-based project 132–33
  'Irmageddon' 139–41
    and speedy data collection 139–40
  light and clarity of the water surveyed (and) 141–43
    potentially limited survey results 141
    power of SDB approach 141
    TCarta contract *and* findings 142–43
  one data set, multiple uses 131–32
  order from chaos (and) 136–38
    autonomous vessels 137–38
    Hurricane Florence, response to 136 *see also* United States (US)
    hydrographic surveys 136–37
    MBEs 137
  ports, boundaries and offshore developments 133–35 *see also* ports
  safety blanket 138–39 *see also* research
  Seabed 2030 project 129–30 *see also* GEFCO
  setting the future course (and) 143–44
    donations of data 144
    Seabed 2030 project 143–44
  from theory to action (and) 135–36
    in Antwerp: trial of *Echodrone* 136

hydrography and bathymetry (and) (*continued*)
    ASV Global: deployment of C-Worker 7 autonomous vessel 135
    SeaRobotics deliver autonomous, unmanned surface vessels 135

illegal fuel smuggling 188
    as international problem 96
incubators/accelerators: HATCH 253–54
    see also European Space Agency
India (and) 111 *see also* NAVIC
    Fisheries Deparrment 225
    government investment in safety technology in fishing sector 224
    Hurricane Okchi protests 225
    Indian National Centre for Ocean Information Services: daily fishing advice 226
    Indian Ocean Tuna Commission 189
    Indian Regional Navigation Satellite system 224–25
    Indian Space Research Organization 224
    Kerala State Government approve budget for rollout of NAVIC 225
    satellite-enabled safety technology beacons – left for children to play with 225
    State Disaster Management Authority 225
    tragedy of loss of life at Tamil Nadu 224
Indonesia (and) 50–52, 111
    data 'donation' to Global Fishing Watch 191
    fish farmers in Lake Toba: incidents of diseased fish 113–14, 117
International Association of Antarctica Tour Operators 76
International Association of Independent Tanker Owners 28
International Hydrographic Organization 129
International Maritime Organization (MO)
    2021 deadline for cyber risk management incorporation into ship safety processes 12–13
    action plan on plastics 156
    restriction on shipping fuel oil 38
    targets for reducing emissions 39
International Monetary Fund (IMF) 139
International Renewable Energy Agency (IRENA) 50
    global projections on growth of offshore wind capacity 57

International Year of the Reef (2018) 165
Internet of Things (IoT) 24, 25, 44, 118, 212
    underwater 124
Ireland (and) 160
    floating wind farm project off west coast (2019) 59
    Seagrass Spotter 160
    Sustainable Energy Authority of Ireland and Saipem engineering company 59
Italy: ban on single-use plastics in Tremiti Islands 155
IUUF: illegal, unreported and unregulated fishing 179–80, 191, 192

Jallal, C 35
James Fisher Mimic: data gathering and monitoring technology 19–20
    see also Briddon, M
Japan (and) 59
    earthquake and tsunami (2011) 150
    Kawasaki Kisen Kaisha (K Line) 25
    Nippon Foundation *and* GEBCO 129
Jiang, J 22
Jonan, I (Energy and Mineral Resources Minister, Indonesia) 64

Kao, E 149
Karokaro, A 114
Kenya: iCow agriculture information service 234

Lambert, N (International Director, NLA) 195
Larantuka Strait, tidal range in 54–55
Leblanc, L (President, Fisheries Safety Association of Nova Scotia) 219
Lidar 132–34
    limited to depth of less than 50 metres 133, 134
liquified natural gas (LNG) 74–75, 76
    and AIDA Cruises: the world's first cruise ship powered by LNG 74
Lloyd's Register/Lloyd's Register foundation 24
Lutoff-Perlo, L (President and CEO, Celebrity Cruises) 72

MacArthur, Dame E 155
    and record for fastest solo circumnavigation of the globe (2005) 155
McCue, D 60
McDermott (offshore engineering group) 20

INDEX

machine learning 14–15, 22, 48, 81, 84, 87, 97, 99, 100, 107, 114, 116, 118, 124, 161, 183–84, 186–88, 257
Madam, S (CEO, TankerTrackers.com) 98
MAERSK 150, 168
Maersk (and) 24 *see also* cyber attacks
  Line 8
  Tankers 17
  wind power/sail technology 17
Manolin Team (Bergen) and notification of sea-lice outbreaks 114–15
marine accidents and machinery failure 23
marine gas oil (MGO), Hapag-Lloyd cruise expeditions using 75
Marine Protected Areas (MPAs) xi, 162, 163
Maritime Resilience and Integrity of Navigation (MarRINav) project 11
maritime surveillance (and) 89–108 *see also* autonomous and remotely operated vessels
  dark matters 97–99
  digital dodging: maritime crime 100–101
  maintaining eyes on the sea 89–93
  oil spills, identifying from space 99–100 *see also* EASOS
  quality, quantity and safety of data (and) 103–06
    avoidance of detection 104
    removal of harm to human operatives 104–05
    robocams *and* USVs operating by stealth 102–03
  SAR-struck 93–97
MARPOL standards (International Convention for Prevention of Pollution from Ships) 40
Martin, E 12
Mason, G (Chief Financial Officer, BP) 9
Mathisen, M 73
Medici, A 219
Mexico and super MPA 162
mobile phone data (anonymized) made available for social development projects 190–91
Modus Seabed Intervention (and) 62–63
  Subsea AVISIoN project 62–63, 64
  trial of innovative AUV docking station National Renewable Energy Centre 62–63
  unmanned underwater vehicle systems 62–63
Monterey Bay National Marine Sanctuary 157–58

and the Marine Mammal Protection Act 157–58
Morocco se Atlan Space drones
Mowi 121–22
  and Aquastorm project 121–22
MS *Raold Amundsen* (owned by Hurtigruten) 76
Mugge, R 204
multi-beam echo-sounders (MBESs) 133, 134, 137
  and MBES-enabled autonomous vessels 145

Nahigyan, P 39
National Oceanic and Atmospheric Administration (NOAA) (and)
  data on wrecks 205
  hydrograhic survey in North and South Carolinas after Hurricane Florence 139
  modelling of risk factors of wreck sites 215
  public access to a crowdsourced bathymetry database 144
  watch list 205–06
NAVIC (Indian Regional Navigation Satellite System) 224
Netherlands 62
  and offshore wind market 55
Neves, D 152
New Zealand Precision Farming for Aquaculture project 118–19
NGOs 247, 248
  and funding 237
NLA International Ltd (NLAI) x, 11–12
  consortium 12
  and system to understand impact of environment on stock and vice versa 118
North Korea, sanctions on 98
Norway (and) 121
  JET Seafood tools to digitalize fish trading process 178–79
  Norwegian Maritime Authority and automatic docking 35
Nusa Lembongan (island) 53–55

Ocean Basemap service (Esri) 144
Ocean Cleanup System 150–51
  and scepticism re its success and its potential for harm 151

ocean conservation (and) 147–72 *see also* plastic waste and pollution
  a broader challenge to ecosystems (and) 149–51
    marine species transported on debris 149–50
    Ocean Cleanup System 150–51
  going with the flow (and) 156–58 *see also* Monterey Bay National Marine Sanctuary
    airborne drones 157–58
    autonomous ocean plastic collection devices 157
    buoys fitted with tracking devices 156
    the WasteShark 157
  harnessing the social network 153–56
  micro waves 151–53
  power to (and from) the people (and) 158–60
    'Genes in Space' (Cancer Research UK) 159
    Global Ghost Gear Initiative *and* Reporter app 160
    mobile phones used to map tides – 'Catch the King' campaign 159–60
    Seagrass Spotter app 160
  protecting and preserving the future of seas and oceans 147–49
  pushing the boundaries 162–65
  sounding off 161–62
    and Orcasound project 161
  using corals to halt reef degradation (and) 165–66
    AI and autonomous systems 165–66
    crown-of-thorns starfish (COTS) 166
    preventive measures against coral bleaching 166 *see also* Australia
The Ocean Economy in 2030 1
Ocean Infinity
  AUVs 196, 197, 199, 200
  flagship: *Seabed Constructor* 196
OceanMedallion™: built by Carnival Corporation, now 'Medallion Class' 80–82
OceanMind (and) 94–96, 184–89, 191
  automated alerts 188
  data 184–85
  as data agnostic – a key strength 185
  focus on illegal activity 188
  human experts: fisheries analysts 187
  identifies unofficial ports 185–86
  mapping fishing activity 188
  tracks vessel speeds 186
  unlicensed fishing in areas of interest 188–89
O'Dowd, P 157
offshore renewables (and) 48–70
  areas of greatest potential for wave power 52–53
  channelling the power of the oceans 48–50
  a climate of fear 50–55 *see also* articles/papers
  continuous operation 62–63 *see also* Modus Seabed Intervention
  cross-cutting value *and* impact on employment 66–67
  floating a new idea 58–59
  maintenance from afar 59–62 *see also* studies
  'on land' at sea *and* robotics 63–65
  potential downsides 65–66
    negative effects of offshore wind projects on bird populations 65
  winds of change 55–58
offshore wind, UK and Ireland leading the way in 59
offshore wind markets 55
Oil Companies International Marine Forum 28

Paakkinen, A and H (co-founders of Penguin wave-converting device) 53
Pacific Light Cable Network linking California to points in Southeast Asia 212
Pacific 'Ring of Fire' 51
Padger, J (Chief Experience and Innovation Officer, Carnival Corp) 83
Pai, L 75
Pancasila-Palmerah Bridge link island of Adonara with Flores 54
Papua New Guinea: Integrated Fisheries Management System 179
Patria, N 64
Patriot (Russian-flagged tanker) 98
Penguin wave-converting device 53 *see also* Paakkinen, A and H
Pichai, S (Google CEO) 15
Piukala, P P (Director, Tonga Cable) 211
plastic (as)
  bags, drop in use of 155
  cause of whale's death 147
  danger to marine animals 147–48

in Hong Kong river 148–49
single-use 155
waste and pollution in rivers and seas 147–50
Plunkett, O (CEO, Ocean Infinity) 199
Port of Rotterdam (and)
blockchain technology 42–43
climate-friendly shipping (SAFETY4SEA) scheme 43
innovation programmes and high-level partnerships 42
Internet of Things platform 44
PortXL concept 43
ports
Antwerp *and* trial of *Echodrone* (autonomous survey boat) 136
Payra Deep Sea Port (southern Bangladesh, Barisal Division) 133–34
Port of Gothenburg 41
ports and harbours (and) 31–47 *see also* Abu Dhabi Ports Company; Port of Rotterdam; ports *and* Smart Port City movement
air pollution 36–40
and related health issues and deaths 36–37 *see also* European Union
autonomous systems/vessels 34
leading the global charge against harmful emissions 31–32
ports in a storm: security and risk factors 32–34 *see also* drones
underwater threat 33–34
safe berth *and* automatic docking 35–36
setting the scene for innovation 42–44
taking down the particulates (and) 40–42
Green Sea Guard system 40, 41–42
MARPOL standards 40
penalties for violations of rules 40–41
predictive condition monitoring, need for 22
predictive modelling 160, 213–14
HiLo tool for accident prevention in shipping 23–24
used in New York (2011) to pinpoint fire-risk buildings 26
Pronzati, L (Chief Business Innovation Officer, MSC Cruises) 82–83
Pudwell, S 79
Pulitzer, J 167
and *The New York World* 167
Putin, President V 11

RangeGuard system (Guidance Marine) 222–23
ransomware cyber attack (Port of San Diego, 2018) 9
Rees, M 33
references (for)
aquaculture 109–28
the Blue Economy: introduction 5
conclusion 258
the cruise industry 88
hydrography and bathymetry 146
maritime surveillance 108
ocean conservation 169–72
offshore renewables 69–70
ports and harbours 45–47
safety of life at sea 238
shipping 28–30
sub-sea monitoring 217
sustainable fisheries 193
Regana, D G 219
Reiger, S 78
remotely operated vehicles (ROVs) 62, 197
renewable energy: interview for *Jakarta Post* 64 *see also* Jonan, I *and* Patria, N
Renewable UK and Floating Wind Action Group 68
reports (on)
26% increase in marine acidity levels since mid-18thC (UN Sustainable Development Goals, 2018 update) 164
ability to predict failure of subsea cables (Offshore Renewables Catapult, 2018) 213
cutting offshore windsector costs (WindEurope, 2018) 67
effect of 2010 Deepwater Horizon catastrophe on submerged wrecks (*Frontiers in Marine Science*, 2019) 204
European policy blueprint on enabling offshore floating wind power costs to fall (WindEurope, 2018) 64–65
Global Maritime Issues Monitor (2018) 7–8, 13
human rights abuses and forced labour widespread in Thailand's fishing industry (Human Rights Watch, 2018) 227
ignoring of safety procedures within ports (Seafarers UK, 2018) 226

reports (on) (*continued*)
  importance of subsea cable transfer to effectiveness of offshore renewables market (Genillard and Co, 2017) 213
  increase in global cruise fleet of 22 percent over the next decade (*Cruise Industry News, Annual Report* (2018–19) 73
  industrial marine acqaculture market: Ocean economy report (OECD, 2016) 110
  likely benefit to South Carolina of increase in offshore wind-related jobs 66
  microplastics found in fish caught off Portuguese coast (*Marine Pollution Bulletin*, 2015) 152
  microplastics ingested by organisms in Mariana trench and other ocean areas (*Royal Society Open Science* journal, 2019) 152
  Norwegian aquaculture PricewaterhouseCoopes, 2017) 123
  offshore wind Carnegie Institution for Science (Stanford, California) 57–58
  public belief in serious threat to marine environment from human activities (*Ocean and Coastal Management*, January 2018) 161
  risks of disruption associated with GNSS (UK Blackett Report *and* London Economics report, 2017) 11
  senior maritime stakeholders on industry and global issues (2018) 7 *see also* Chambers, S
research (on)
  abundance of fish population in Cabo Pulmo National Park (Mexico) 163
  aquaculture using space-based Earth observation, machine learning *and* underwater Internet of Things (NLA) 124
  consistently low and decreasing accident rate in commercial aviation (Boeing) 202
  economic impacts of increasing sea-level rise 139 *see also* Ruiz-Ramirez, J
  fatalities in the fishing industry : 62 per 100,000 workers (UK Maritime and Coastguard Agency, July 2018) 223
  longer-term predictions of fish abundance (CSIR-National Institute of Oceanography, India) 226
  microplastics and corals (*Marine Pollution Bulletin*, Duke University, 2017) 152
  plastic in river systems as source of 95% of plastic entering oceans (*Environmental Science and Technology*) 149
  underwater lighting systems (Centre for Environment, Fisheries and Aquaculture Science) 177 *see also* SafetyNet Technologies
Rieger, S 78
robotics
  ANYbiotics 64
  ANYmal 64
  SeaRobotics and autonomous unmanned surface vessels 135
robots
  involved in repairing cables 211 except in depths with high water pressure 212
  Racemaster 3.0 (neat-cleaning) 120–21
Rodgers, P (CEO, Euronav) 9
Rolls-Royce 22–23
  and autonomous deep sea cargo ships 22–23
Rose, J 227
Rotterdam, Port of
  Environmental Ship Index 39–40
  financial incentives for low- or zero-carbon vessels 39–40
  targets for reducing carbon dioxide emissions 39
Ruiz-Ramirez, J 139
Rushlight Environmental Analysis and Metrology Award 40 *see also* Green Sea Guard system
Russell, K 91
Russia
  blamed for attempted disruption of GPS systems 11
  and ties to fuel smuggling 98

Sadlier, G 11
safety of life at sea (and) 218–38
  caught in the net: forced, migrant and slave labour 227–28 *see also* reports
  cruise control (and) 228–31 *see also* safety at sea tracking devices
    CCTV technology for assisting in man overboard (MOB) incidents 229

MOB detection system and other tracking devices 230–31
passengers falling overboard *and* average number of incidents 228–29
perception of safety as important factor for cruise companies 229
forewarned is forearmed 226 *see also* research
funding the future, sources of 233
invest to save (and) 234–35
   companies investing 235
   iCow agriculture information service 234
   Omidyar network 235
   SeeClickFix 234–35
   using apps 234–35
no safety net 223–26 *see also* fishing fatalities *and* research
proof positive 232–33
protecting human life in harsh environments 218–21
   and selfie-related deaths 220
VR, wearables and robots to the rescue 221–23
   RangeGuard system for reducing risk of collision 222–23
   VR techniques in training situations 221–22
   wearable bracelets (MSC Cruises) 222
Safety of Life at Sea Convention 132 *see also* hydrography and bathymetry
safety at sea tracking devices
   McMurdo Smartfind S10: an AIS MOB device (McMurdo Marine) 230
   MOB detection systems 230
   MOBtronic device (MARSS Group, Monaco and London) 231
   'Best Safety Product of the Year', IHS Markit Safety at Sea Awards 231
   Sea-Marshall AU9 (SeaSafe Systems, UK) 230
SafetyNet Technologies (and) 176–77
   testing of systems in partnership with Young's Seafood 176–77 *see also* research
   underwater lighting systems 176
Salazar, M 119
Salerno, B 92 *see also* articles/papers
satellite(s) (and)
   data used by intelligence agencies investigating fuel smuggling in Yellow Sea 98
   Digital Globe Earth observation 143
   NovaSAR-1 93
   operated by Surrey Satellite Technology Limited (UK) 93–94
   SAR date 187
   ZACube-2 ('Africa's most advanced nanosatellite') 93
satellite-derived bathymetry (SDB) 59, 113, 140–41, 145–46, 191
   major advantage of 143
   technique 140
Scotland (and) 121
   acqaculture industry in Shetland *and* HAB Report 117
   the Blue Economy 111–12
   Hywind Scotland floating wind farm 58–59
   Scottish Centre of Excellence in Satellite Applications 118
SDB *see* satellite-derived bathymetry (SDB)
'sea blindness' phenomenon xi–xii
Second World War battlefields 247
Secretariat of the Pacific Regional Environment Programme (SPREP) 205
selfie-related deaths 220
   and 'no-selfie zones' in Mumbai 220
Setch, E (saved from the sea) 220
Shell 23–24
   fish farms 116
   and Hilo predictive modelling tool 23
Shi Wei, N 34
shipping (and) 6–30
   innovation options – electric propulsion *and* wind power 16–17
   navigating an unprecedented sea of data 6–9 *see also* reports
   playing 'spoof' (and) 10–12 *see also* studies
      GPS signals 10
      MarRINav project 11 *see also* NLA International Ltd
   the power of sharing 23–24 *see also* HiLo
   from reaction to prediction (and) 17–23
      condition-based monitoring 18–21
      need for predictive condition monitoring 22
      reliability centred maintenance (RCM) 17–18
   rising to the cyber challenge 12–13
      see also KPMG; Kongsberg *and* Wärtsila
   tackling the energy challenge 13–16
      see also Carlsson, L *and* Pichai, S
      and ship design 14

Shukla, F 164
Singapore (and)
    development of autonomous vessels 34
    Keppel Offshore & Marine 34
    Maritime and Port Authority of
        Singapore 34
    PortXL concept 43
    Technology Centre for Offshore and
        Marine 34
single-use plastics: banned in Trimiti Islands,
    Italy 155
Slat, B (Dutch marine entrepreneur) 168
Small Island Developing States (SIDs) 139,
    140, 205
Smart Port City movement 32
    Dutch–Belgian project and the
        Internet of Things/'Internet of
        Tarpaulins' 32
smartphone revolution 212
Smith, J (Maritime Coordinator,
    International Transport Workers'
    Federation) 219
smuggling
    as critical resource for Philippine
        Government 96–97
    of fuel in the Yellow Sea (trade route
        between Korea and China) 98
Southeast Asia (and) 7, 139, 212
    acquaculture 110–11
    cable failures 210
    energy consumption 49–50
Stafford, R 180
Standard Supporter supply ship,
    loss of 62
Stephanos, A 218
Stephanos, B 218
Streisand, B 162
studies (on)
    disruptions of GPSs (Centre for
        Advanced Defense, 2019) 11
        see also Edwards, J and Russia
    invasive rats as major threat to tropical
        island coral reefs (2018) 164
    over-fishing causing removal of natural
        predators of coral-eating snails
        (2018) 164
    plastic pollution and increase of
        likelihood of disease in corals
        (*Science*, 2018) 152
    plastic waste entering ocean from 192
        countries (*Science*, 2015) 149
    plastic waste entering sea from
        Shing Mun River, Hong Kong
        (Greenpeace, 2018) 148–49
    small-scale fish traceability (Thai Union,
        USAID and Mars-Petcare) 180–81
    underwater inspection methods research
        (2017) Carbon Trust Offshore
        Wind Accelerator 60
Suastika, K (Chief Commander, local police,
    Nusa Lembongan) 220
sub-sea monitoring (and/and the) 194–217
    continual challenge 199–200
    establishing the search rhythm 198–99
    fragility of the connected world
        (and) 211–13
        broken cable between Tonga and
            Fiji 211
        getting power back to shore 213–15
        see also reports
    loss of flight MH370 (2014)
        (and) 202–04 see also research
        and reports
        danger from oil spilled during
            Deepwater Horizon catastrophe
            (2010) 204
        potential danger from WW2 sunken
            ships 203–04
        rarity of civilian airline crashes over
            water/deep water crashes 202–03
        recorded fatal accidents 203
    needles in a watery haystack
        (and) 201–02
        ship losses (2017) 201
        submarine risk levels 202
    new possibilities 206–07
    raising the profile of dangerous
        wrecks 205–06
    safeguarding the world's data
        (and) 209–10
        Dunant cable (Google) 210
        first transatlantic cable 209–10
            see also Buchanan, J and Victoria,
            Queen
    seabed constructor 196–97
    shining a light on mysteries of the
        deep 194–96
    'system of systems' (and) 207–09
        challenge of returning collected data
            subsurface to shore 208
        ease of launching unmanned and
            autonomous vessels 208–09
        USVs as navigation aids for
            AUVs 208
survey of Cayman Lidar 133 see also multi-
    beam echo-sounders (MBEs)
sustainable fisheries (and) 173–93 see also
    IUUF

airborne AI 182–89 *see also* Atlan Space; drones *and* Ocean Mind
casting a digital net to help feed the world 173–76
a light in the blue 176–77 *see also* SafetyNet Technologies
a multi-faceted problem (and) 179–82 *see also* United Nations
    tackling IUUF 179, 180 *see also* United States
    Thai Union 180–81
    ThisFish (Vancouver) and similar pilot programmes 181
the open data challenge 189–92
providing the framework for 177–79 *see also* Norway: Papua New Guinea *and* United States (US)
Synthetic Aperture Radar imagery 186

Tan, A 212
TCarta awarded Antigua and Barbuda contract (2017) 142
tech-enabled wearables, eg. MSC Cruises' wearable bracelet 222
Telstra (and) 214–15, 216
    30% reduction in service impacts on subsea cables network 214–15
    community engagement to counter threat from bottom-trawling fishing 214–15
    investment in better use of available data to help plan cable routes 215
terrorism/terrorist groups 33 *see also* drones
Thai Union 180–81 *see also* studies
Thailand and potential for slavery in its fishing industry 227 *see also* reports
Tonga to Fiji subsea cable 211 *see also* Asian Development Bank *and* World Bank
Tongan Government and internet connectivity 216
Tozzi, J 79

UNESCO Oceanographic Commission 129
United Kingdom (UK) (and)
    air pollution in 37
    Centre of Environment, Fisheries and Aquaculture 11, 140
    Collaborative Centre for Sustainble Aquaculture Futures 111
    Conflict Stability and Security Fund 132
    Defence Attaché to Buenos Aires 196
    Fishermen's Mission 223
    Floating Wind Action Group 68
    Gas and Oil Authority project on transition and synergy potential of offshore wind 67
    Government Offshore Wind SectorL expected increase in 'green collar' jobs in 66
    Greater London Authority and open data hubs 189
    Hydrographic Office and bathymetry-based project: Cayman Islands (2018) 132
    London Data Store 189
    National Data Centre 196
    Offshore Renewables Catapult 213, 216 *see also* reports
    Offshore Wind Industry Council 56
    Offshore Wind Industry Prospectus (2018) 56
    offshore wind market 55
    Open Data Institute 43–44, 189
    ports sector 31–32
    Satellite Applications Catapult 99
    Submarine Parachute Group of the Royal Navy 196
United Nations (UN) 2
    Conference on Trade and Development on operation of merchant ships as percentage of world trade 6
    Global Pulse programme 190
    and growth of world population 2
    Ocean Conference (New York, 2017) 143
    Sustainable Development Goal (SDGs) 14, 50, 178
        Action 14.4 – ending IUUF 179
        target 163
United States (US) (and)
    AliMoSphere project (California) 158 *see also* Gaworecki, M
    Business Network for Offshore Wind (Maryland) 67
    Californian Air Pollution Control Fund 41
    Coast Guard: responsibility for local area contingency plans for spills 205
    global energy demand in North America 49
    National Oceanic and Atmospheric Administration (and)
        acquaculture research 122
        Office of Coast Survey navigation response team and Hurricane Florence 136
    Navy 196
    oil export ban on Iran (2018) 97
    Port of Los Angeles

United States (US) (and) (*continued*)
    Clean Air Action Plan (2023) 39
    Inventory of Air Emissions (2017) 38–39
    predictions of green collar jobs in 66
    State of Hawaii and Territory of American Samoa partnership with National Oceanic and Atmospheric Administration and Liquid Robotics (Boeing company) 163
    and Wave Glider USV 163
    West Coast 59
unmanned aerial vehicles (UAVs) *see* drones
USS *Thresher* and USS *Scorpion* lost at sea (1963 and 1968 respectively) 198, 200

Vaughan, A 48
Venkataramanan, K 33
Victoria, HM Queen 209, 210
Vietnam (and)
    Association of Seafood Exporters and Producers: investment in developing aquaculture industry 111
    intention to become world's top aquaculture producer 110–11
Viking and Flettnor rotor 77 *see also* Flettner, A
Viking Line cruise ship: *Viking Grace* 76–77
    and Norsepower rotor sails 76–77
virtual reality (VR) techniques in training situations 221–22
    for offshore wind sector (Newcastle College, UK) 222
Vleugels, A 32
Vodafone/Vodafone Foundation 190, 191, 193, 233

Wall, R 16
Wanchang, L 220
Wärtsila 12
Watts, J 152
wave-converting/wave-level devices 53–54
Wello Oy and Penguin wave energy converter 53–56
Westfield 78–79, 87, 190
White, C 110
Williams, A 219

Wilson, B 92
wind farms
    and AVISIoN 62
    Block Island (offshore) 56–57
    offshore 56–58, 60–63, 65, 134, 213, 214, 222
    floating 58–59
wind power 16–17, 57, 59, 64–65, 76–77
    and new offshore installations across 9 markets globally (2017) 55
WindEurope 64–65 *see also* reports
    proactive stance on issue of port infrastructure 67
Windward Ltd: use of satellite data to track movements of Iranian oil tankers 97–98
Winfrey, O 153
Wingrove, M 20, 86
World Bank (and)
    estimate of cost of poor management/over-exploitation of fishery resources 174
    estimate of size/scale of growth of acquaculture market (2014) 109
    funding of sub-sea cable (jointly with Asian Development Bank) 211
    heavy investment in development of geothermal energy in Indonesia 51
    portal of open data 190, 193
World Health Organization, air pollution limits set by 37
World Shipping Council 28
World Wide Fund for Nature (WWF) 1–2
    assessment of key ocean assets 2
wrecks
    dangerous 205–06
    long-lost 198
    marking sites of 34
    over the centuries 203
    in Pacific region (approximately 10% being oil tankers) 204
    risk from 215–17
    Second World War 203–04, 247
Wyatt, T 150

Yousafzai, M (Nobel Prize Winner) 72

Zachos, E 148